# Label Embellishments and Special Applications

Exploring the techniques and processes used for adding decorative finishes and functionality to labels

# Other Labels & Labeling books:

**ENCYCLOPEDIA OF LABEL TECHNOLOGY**
Michael Fairley

**THE HISTORY OF LABELS**
Michael Fairley and Tony White

**DIGITAL LABEL AND PACKAGE PRINTING**
Michael Fairley

**ENVIRONMENTAL PERFORMANCE AND SUSTAINABLE LABELING**
Michael Fairley and Danielle Jerschefske

**CONVENTIONAL LABEL PRINTING PROCESSES**
John Morton and Robert Shimmin

**LABEL DESIGN AND ORIGINATION**
John Morton and Robert Shimmin

**LABEL DISPENSING AND APPLICATION TECHNOLOGY**
Michael Fairley

**CODES AND CODING TECHNOLOGY**
Michael Fairley

**LABEL EMBELLISHMENTS AND SPECIAL APPLICATIONS**
John Morton and Robert Shimmin

**BRAND PROTECTION, SECURITY LABELING AND PACKAGING**
Jeremy Plimmer

**DIE-CUTTING AND TOOLING**
Michael Fairley

**MANAGEMENT INFORMATION SYSTEMS AND WORKFLOW AUTOMATION**
Michael Fairley

**SHRINK SLEEVE TECHNOLOGY**
Michael Fairley and Séamus Lafferty

**LABEL MARKETS AND APPLICATIONS**
John Penhallow

For the latest list please visit: **www.labelsandlabeling.com**

# Label Embellishments and Special Applications

Exploring the techniques and processes used for adding decorative finishes and functionality to labels

**John Morton and Robert Shimmin**
4impression

**Label Embellishments and Special Applications**

Exploring the techniques and processes used for adding decorative finishes and functionality to labels

First edition published 2016 by:
Tarsus Exhibitions & Publishing Ltd

Printed by CreateSpace, an Amazon.com company.

ISBN 978-1-910507-08-7

# Contents

While every care has been taken to ensure the information, charts, diagrams and illustrations in this publication are correct at the time of publishing it is possible that technology, specifications, markets and applications, or terminology may change at any time, or that the author's or contributor's research or interpretation may not be regarded as the latest accepted guidance in some parts of the world of labels.

The publishers therefore cannot accept responsibility for any errors of interpretation or for any actions, decisions or practices that readers may take based on the publication content and would advise that the latest industry supplier specifications, standards, legislative requirements, performance guidelines, practices and methodology should always be sought before any investment or implementation is made.

# Foreword

The label today is more than just a simple conveyor of product information.

Brands competing for the attention of the consumer continue to strive to make themselves more visually appealing on the retail shelf by adopting a wide range of embellishment and decoration techniques.

This book will explain in detail the more traditional methods of label enhancement using foiling, embossing and novel printing techniques, but it will give equal weight to the new technologies and innovations that are allowing packaging to engage a wider range of human senses.

The use of smart materials, developments in multi-web manufacturing and the introduction of electronics in particular, are accelerating the evolution of the label as a more functional packaging component.

This book aims to provide an insight into the wide range of processes, techniques and developments that continue to add new dimensions to the once humble label.

**John Morton and Robert Shimmin**
4impression Limited

# About the Label Academy

This book is part of the recommended study material for the Label Academy, a global training and certification program for the label industry. The Label Academy was created by the team behind Labels & Labeling magazine and the Labelexpo series of events.

The Academy consists of a series of self-study modules, combining free access to relevant articles and videos with paid text books (both printed and electronic). Once a student has completed a module, there is an opportunity to take an online test and earn a certificate.

It is expected that a Label Academy qualification will become a standard in the industry – for printers/converters, suppliers, brand owners and designers – and assist in providing a benchmark. In addition to its own training, the Label Academy will aim to become a resource provider to the many existing educational programs in the industry. Accredited training courses will be promoted through the Label Academy website and books will be provided at discounted rates.

The Label Academy concept was pioneered by industry expert Mike Fairley. This was in response to a reduction in the number of dedicated printing colleges and the need to standardize training across the world. The label industry also has its own specific training needs – it has some of the widest range of materials, printing processes and finishing solutions of any printing sector.

We are also working with other training experts and authors to ensure that the Label Academy provides up-to-date and relevant training material for the industry.

The Label Academy is supported by the key trade associations, including FINAT, TLMI and the LMAI.

**www.label-academy.com**

# Label Academy sponsors

Thank you to our founding sponsors, without whom this ambitious project would not have been possible:

### Cerm

Cerm designs business automation software solutions to meet the specific demands of flexo and digital narrow web printers. Using the latest technology, our team's focus is on innovation and continuous improvement.

Our automation solutions support each step in the printer's integrated workflow – from estimating to production, shipment and data collection – and provide the feature and functionality printers need to gain efficiency and improve profitability.

Cerm inspires collaboration and helps printers remain competitive in the market and deliver the best products possible. We are proud to sponsor the Label Academy and contribute to the future of the narrow web printing industry.

**www.cerm.net**

### Flint Group Narrow Web

Flint Group Narrow Web has the products, the solutions, and the technical experts to handle any print situation. Providing solutions for food packaging, sustainability, increased bottom line, efficiency, and uptime – delivering the basics needed to run a successful operation, and the expertise to go above and beyond to another level of success.

Our experts provide solutions to your printing problems with the innovative products and services that have made us an industry leader around the world. Wherever you are, we are – available to help you reach your business goals today and into the future.

Continuous improvement is paramount to Flint Group; we are proud to sponsor the Label Academy and the benefits it will bring to the future of our industry.

**www.flintgrp.com**

## Gallus Group

The Gallus Group with its production sites in Switzerland and Germany is a leader in the development, production and sale of narrow-web, reel-fed presses designed for label manufacturers. The machine portfolio is augmented by a broad range of screen printing plates (Gallus Screeny), globally decentralized service operations, and a broad offering of printing accessories and replacement parts. The comprehensive portfolio also includes consulting services provided by label experts in all relevant printing and process engineering tasks. The Gallus Group is a member of the Heidelberg Group and employs around 430 people, of whom 253 are based in Switzerland. The group headquarters is in St.Gallen, Switzerland.

**www.gallus-group.com**

## MPS Systems B.V.

Producing high-quality label printing depends on several factors; one of them is the operator of the press.

As a press machine builder since 1996, MPS Systems B.V. knows how important training and education on subjects like pre-press, label printing and finishing is. For label printers, it is critical that their operators keep up with pre-press and press developments in addition to label trends. Therefore, MPS sponsors the Label Academy, to advance operator's passion for printing, share expertise and help multiply benefits.

The MPS slogans of 'Printers First' and 'Technology with Respect' have always underlined the core philosophy of MPS from press design to operator satisfaction. We develop our presses with a strong focus on user-friendliness and respect for the press operator: Printers First.

**www.mps4u.com**

## HP Indigo

HP Indigo is a global leader in digital printing, with a broad portfolio of digital presses and workflow solutions. Indigo's proprietary Liquid Electrophotography (LEP) technology delivers exceptional print quality for the widest variety of applications including labels, flexible packaging, shrink sleeves and folding cartons. HP Indigo's digital presses match gravure print quality satisfying the most demanding brands.

A division of HP Inc.'s Graphics Solutions Business, Indigo serves customers in more than 122 countries, including many of the top label and packaging converters worldwide.

**www.hp.com/go/labelsandpackaging**

## UPM Raflatac

In a little more than three decades, UPM Raflatac has become one of the world's leading manufacturers of pressure sensitive label materials, developing and leveraging the latest innovations in adhesive technology. Our film and paper label stocks are used for product and information labeling across a wide range of end-uses – from pharmaceuticals and security to food and beverage applications.

We are an engineering driven company with industry-leading products known for their consistent high quality and top performance. We are also known for the high performing supply chain and undisputed leadership in the area of sustainability. UPM Raflatac's dedication to innovation, sustainability and top quality is matched only by our commitment to service excellence. We call it the Raflatouch.

**www.upmraflatac.com**

# About the authors

**4impression Training**

4impression are specialist providers of training across a wide range of print and packaging related subjects. Staffed by industry trained tutors and supported by a network of print and packaging suppliers, the company delivers face to face to courses providing understanding of print processes, embellishments, materials, origination and finishing. Recently 4impression wrote the FINAT Educational Handbook which covers all aspects of self-adhesive label manufacture. They have also produced a comprehensive range of learning resources for the FINAT Knowledge Hub.

As authors of this book 4impression are uniquely positioned to offer additional personalised training to readers who require more insight into its content. The directors of 4impression, colleagues from their days working for the Jarvis Porter Group, are passionate about print and have a long track record in delivering courses to major packaging users and their supply chains.

**John Morton**

John has hands-on experience of all the major printing processes and has held operational and technical development roles at director level in the packaging sector. A qualified printer, John's career spans magazine production, commercial print, packaging and label production. Before joining 4impression John was actively involved in the Unilever advanced printing and decoration training courses attended by delegates from operations around the globe.

**Robert Shimmin**

Robert has held senior marketing and business development positions in the print, packaging and label sector spanning more than 20 years. He is a regular contributor of articles to the print and packaging trade press and has supported initiatives that seek to build awareness of the latest research and innovations emanating from UK universities. In addition to his involvement with 4impression he runs Shimmin Associates, a research and marketing consultancy offering support to both UK and international customers in the label and packaging sector.

**Paul Jarvis**

Paul Jarvis, formerly chairman of Jarvis Porter Group PLC, oversaw its growth to become one of Europe's leading packaging suppliers with a turnover in excess of £100 million, employing 1,600 people in 7 countries including the United States. Paul was a director and founder member of the Leeds Training and Enterprise Council and represented CCL Label on the main board of FINAT (the world-wide association for self-adhesive labels and related products). Paul provides strategic direction to the packaging and print sectors capitalising on his vast experience and global network of contacts.

www.4impression.com

# Acknowledgements

# Chapter 1

## Exploring the opportunities to add value to labels

This book explores the techniques and processes used for adding decorative finishes, special effects and functionality to labels.

The desire to embellish printed material with a luxury or metallic look can be traced back over hundreds of years, from the decoration of the earliest printed books in the 15th century with gold leaf. Since then the development of the bronzing process and then metallic inks, up to the latest innovations in hot and cold stamping foils have given brand owners more options in their quest, to differentiate their products. The early chapters in this Handbook track the evolution of techniques that will add a decorative dimension to the surface of the label.

More recently innovations in printing materials and techniques have unleashed a new breed of smart or intelligent products that can not only enhance the visual appearance of the label, but can also offer specific functionalities. In some cases these innovations allow the label to appeal to a wider range of human senses and even interact with changing environmental conditions.

Finally this book will cover the evolution of processes that facilitate the creation of multi-layer labels and booklets and allow print and embellishments to be applied not just to a single surface, but to a range of surfaces, perhaps within a booklet or form style construction.

### WHAT IS AN EMBELLISHMENT?
*Definition of embellishment - to make more beautiful and attractive; to decorate.*

An embellishment adds an additional decorative finish to the surface of a label in order to enhance its visual appeal and make it more eye catching.

Perhaps the most common form of label embellishment is foil-blocking, where parts of the label design can appear in a high gloss reflective foil. Its superb brightness and mirror finish has effectively been used to simulate its historic predecessor, gold leaf. But today foiling is only one of many techniques and processes available to brand owners seeking to add value to their brands

The role of the label as a packaging component is complex and rapidly evolving. From its early beginnings as a simple descriptive marker its role has been transformed. Today the function of the label is more diverse than ever before.

The label has always performed a key role in conveying mandatory and critical information, but the rise of the global brand has elevated the importance of the brand image and shelf-impact to new heights. The role of the label as a marketing tool reflecting the brand owners' carefully created image is critical. In some instances labels have become 'works of art' with a huge number of print processes, materials and embellishments at the disposal of the designer.

## KEY MARKET SECTORS

There are particular market sectors such as the health and beauty care and premium drinks segments where the visual appeal of the branding and packaging is the most important function of the label.

In many other market sectors, premium sub-sectors are emerging as brand owners and retailers attempt to target more affluent consumers.

Here competition between brands and the desire to create visual differentiation on the supermarket shelf are driving an increase in the sophistication of label graphics and in the use of surface decorative embellishments.

In health and beauty care the intensifying competition between proprietary and own label brands is resulting in the introduction of 'own label' premium lines and a trend towards more innovative and visually striking labels.

The health and beauty care segment has one of the highest penetration levels of self-adhesive labeling. In this sector filmic substrates that are resistant to moisture are heavily used and the use of clear PP films to create a 'no-label' look, is popular.

Surface embellishments used on filmic materials offer their own particular challenges and there has been a trend towards tactile finishes that improve product handling performance for the user in wet conditions.

Within the drinks market the high value spirits sector has always used highly decorative labels, with many incorporating security features such as holograms, security threads, security inks and a variety of other anti-theft devices or counterfeit deterrence features (Figure 1.1).

## EMERGENCE OF NEW PREMIUM SECTORS

New premium sectors are also emerging. One of the most rapidly growing areas of use for self-adhesive labels is premium beers. Here decorative labels with a high added value content printed primarily onto filmic labels are making significant inroads (Figure 1.2). Glue applied labeling technology still dominates for high-

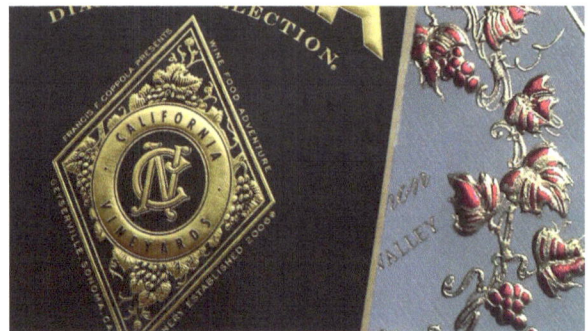

**Figure 1.1 -** Examples of labels with a high decorative content

volume applications in both beers and spirits with many labels having a high decorative content in the form of metallic inks.

Although private food label brands are typically associated with cheap, low quality alternatives to name brands, retailers are now starting to develop their own 'premium' ranges (Figure 1.3). Tesco Finest, for example, uses black and metallic packaging to convey high quality.

To continue to see this upsurge in private label purchases, they are going to have to innovate rather than imitate. Proprietary shapes, textures, colors, imagery and typography leave a lasting impression with consumers and allow products to stand out from the shelf.

## LABELS THAT EXPLOIT A WIDER RANGE OF SENSES

Developments in the decorative function of labels will continue to evolve. A new area of interest in design is

**Figure 1.2 -** Growth in use of filmic labels with high decorative content for new premium beer sectors

**Figure 1.3 -** Premium ranges are emerging in the food sector with high added value labels

exploiting the human senses of touch and smell. We are already seeing labels incorporating surface textures and 'scratch and sniff inks' appearing on the retailer's shelves. Premium packaging is now seeking to engage all of a consumer's senses, with tactile and

olfactory trends emerging. Sight, is the most important sense when it comes to packaging, so we are seeing a move away from poorly used white space, to vibrant, colorful, eye catching designs. The power of touch and rich textures are being exploited and there is a trend towards sustainability and biodegradability.

## OTHER MARKET DRIVERS AND DEVELOPMENTS

There are many other factors that are having an impact on the evolution of label structures.

New marketing and promotional demands that require more information to be carried on-pack, are resulting in new multi-layer structures and booklets that require a different approach to manufacturing and subsequent printing.

Significant escalation in the counterfeiting of goods and packaging and the rapid rise in retail theft has resulted in the emergence of new processes such as holographic foils which have both decorative and security benefits.

An exciting development is the ability to integrate active and intelligent components (such as RFID features) into the label itself.

New inks and materials mean that labels can react to changes in environmental conditions such as heat and temperature by changing color and this means that they can be used as visual indicators of product condition or freshness.

It is clear that the uses of labels as packaging components are many and diverse and will continue to expand. The end user now has the potential to incorporate a variety of decorative and functional features into a single label.

## SUMMARY OF DECORATIVE EMBELLISHMENTS AND SPECIAL APPLICATIONS

A brief summary of all the decorative embellishments and added value features covered in the Handbook are detailed in Figure 1.4.

## COST CONSIDERATIONS

Finally when considering the use of decorative embellishments the following cost considerations

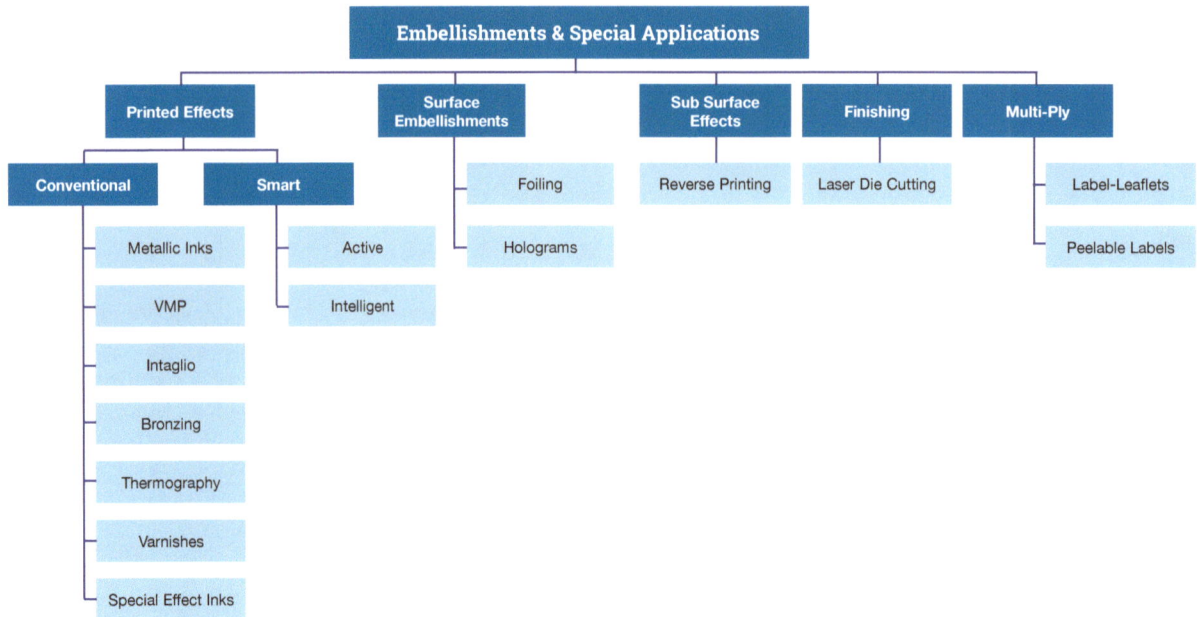

**Figure 1.4 -** Summary of embellishments and added value features

must be evaluated. These factors can impact heavily on project costs, thereby influencing the choice of decoration used for a label or pack.

- Capital cost of equipment – The capital cost of embellishment equipment must be considered
- Materials – Cost of prep, tooling, foil, special inks and adhesive
- Cost of die or plates – Hot foil requires a die whilst cold foil requires ink and photopolymer plates
- Set-up/makeready costs – Increased makeready times dependent on the type of decorating technology
- Press speed – Press speed can be significantly affected by the embellishment process
- Waste factor – Whether foiling or printing; be sure to include the appropriate waste factor

# Chapter 2

# Foil stamping processes

Metallic finishes and embellishments are favored by packaging designers for their ability to create a striking visual addition to a brand's image.

There are a number of processes that have evolved over the years to produce a metallic finish including, foiling, metallic inks and bronzing. Each of these processes will be dealt with in some detail in this book.

Foiling or foil blocking is a dry printing process which uses very thin aluminum foil in a variety of metallic colors – such as gold, silver, red or blue – rather than inks, from which to embellish parts of the label design.

Hot foil stamping is a printing process that uses heat and pressure to transfer the image from a metal printing plate to a label substrate, using a metallic or colored foil to produce the desired result.

A more recent development of foil blocking is the cold foil process, in which a print unit is used to print a special adhesive on the label web where the metallic effect is required. When the metallic foil is brought into contact with the adhesive it adheres to it to produce the printed foil design on the label. A comprehensive insight into the cold foiling process will also be provided later in this chapter.

## HISTORICAL EVOLUTION OF FOILING

Hot foil stamping, a method of transferring a metallic or pigment finish from a carrier strip of paper or film – known as a foil – onto a label substrate using heat, pressure and dwell time, was first patented by Dr Ernest Oeser in 1892. This first hot stamping process

used both gold and colored stamping foils made from 23 carat gold or bronze powder with a dye to obtain the necessary color. Both these types of foils were supported on a glassine carrier strip.

Then, in 1931, the world's first vapor deposit stamping foil using real gold was invented and patented by Konrad Kurz. This was mainly used in industrial applications for the foil printing of pencils, in bookbinding, and in the hat industry.

The early stamping foils had a number of limitations. The use of real gold made the foils very expensive, while the bronze powder foils were easily tarnished. In addition, the coloring dyes used with the metallic powders were not light stable and tended to fade. This meant that other alternatives for metallic image printing continued to be considered.

However, it was not until the 1950s that vacuum or vapor metallized foils using aluminum were developed and introduced by Leonhard Kurz GmbH & Co. This company, founded in the 1890s by Leonhard Kurz as an enterprise for the production and marketing of gold leaf, now had a product that was to revolutionize metallic image printing throughout the graphic arts industry – and is still used today.

In this hot foil stamping process, a pigmented or metalized coated foil was transferred from a carrier, more usually today a polyester film, and fused to a substrate by heat in a hot stamping machine.

This means that the pigmented or metalized coating has to be compatible with the material to be stamped. For this reason, hot stamping foils are manufactured in various formulations designed to give quality prints on a specific material. Also incorporated in the foils are qualities such as abrasion resistance, oil and grease resistance, and chemical resistance.

Over the years an enormous range of foil types has become available. Where metallics are concerned, there is every conceivable shade of gold (and silver), in bright, satin or matt finishes. These are supplemented by metallic colors, brushed finishes and a wide range of gloss and matt colored pigments. In addition, there are numerous patterned colored and metallic foils which are particularly suitable for background effects. To meet more specialized requirements today, there are also fluorescent, magnetic, pearlized and holographic printing foils.

Roll-on stamping machines for applying the foils were introduced by Kurz in the 1960s and vertical stamping machines in the 1970s.

Many other companies have also contributed to the development and growth of metallic image decoration on labels, from the likes of Gallus, Nilpeter, Edale and other early 20th century roll-label press manufacturers, to dedicated hot and cold foil materials companies such as George M Whiley and API Foils.

A more in-depth narrative on the historical evolution of print and embellishing processes can be found in *The History of Labels: The evolution of the label industry in Europe* by Michael Fairley and Tony White.

## HOT FOIL STAMPING

Hot foil stamping is a decorative embellishment which uses heat and pressure to transfer an image from a metal printing plate to a label substrate using metallic foil to give a highly reflective surface. Foil is often gold or silver, but it is also manufactured in various colors, patterns and special finishes (see Figure 2.1). A very

hard thermoformed plastic plate can be used instead of metal for very short runs.

The use of metallic finishes can considerably enhance both the quality and the added value of the label giving excellent solid images, opaque finishes and producing a very high density of color.

**Figure 2.1 -** Example of hot foiling

## THE FOILING PROCESS

A lacquered aluminium foil is positioned with the adhesive layer face down on the substrate. Pressure is applied with a heated "imaged" die to activate the adhesive layer. The foiling film is then separated from the substrate leaving the metallized image of the die reproduced on the substrate surface.

The stamping foil is made up of a number of layers - film carrier - release coating – reflective lacquer and adhesive.

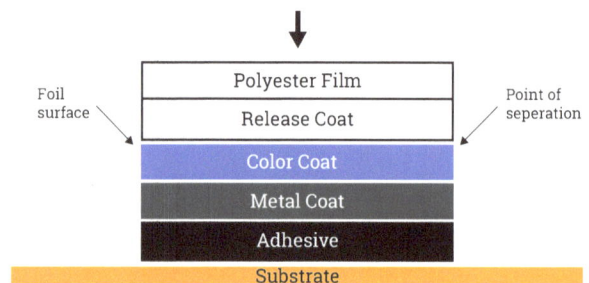

**Figure 2.2 -** Structure and composition of foil

Figure 2.2 illustrates the structure and composition

of the 'foil' when positioned over the substrate to be foil printed and prior to contact with the heated foiling die.

The foil is positioned with the adhesive face down on to the substrate and pressure is applied with a heated engraved die which activates the adhesive. The activation of the adhesive is for a very short period, known as the 'dwell time'.

The control of die temperature and pressure between the substrate, foil and die, is very important in ensuring that the adhesive is activated sufficiently for the adhesive and lacquer film to transfer to the substrate.

For effective transfer to take place consideration must also be given to the type of foil, the release factor of the adhesive and the surface of the substrate being foil blocked.

## METHODS OF FOILING

Hot foil stamping (also known as foil blocking) can be used as a single off-line process or can be incorporated into a multi-process operation, running in-line with other printing and embellishing processes.

Units designed for the off-line hot-foil printing or decoration of labels, come in a variety of configurations and widths. Stand-alone hot foil blocking machines are generally narrow web, but the standard self-adhesive production widths are used where combination print processes include foil blocking. Larger sheet-fed presses are used for foil blocking of large sheets of glue-applied or in-mold labels.

There are three different types of foiling methods each having its own distinct advantages:-

**Flatbed foiling -** offers an easier set-up (make-ready) and tooling costs are significantly less.

**Round flatbed foiling -** is better suited for the shorter runs lengths and uses the same flat die tooling as the flatbed system.

**Semi Rotary and Full Rotary foiling -** is suitable for longer run lengths, gives higher running speeds and is excellent for fine details, but the cost of the tooling is much higher than flatbed.

## FLATBED FOILING

Flatbed foiling machines work on the same principle as the 'platen' printing presses and die-cutting machines (see Chapter 2 of the *Conventional Label Printing Processes book*).

The imaged dies are usually located onto the electrically heated bottom platen using a honeycomb base and held with tensioning clips. These clips allow the operator to make positional adjustments to the die to achieve the correct print registration (see Figure 2.3).

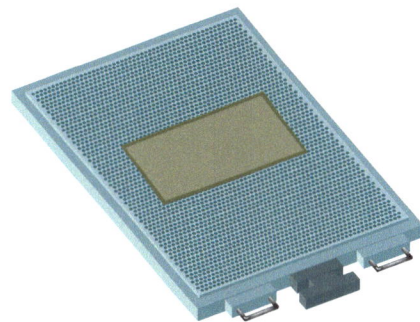

**Figure 2.3 -** Honeycomb base with embossing plate mounted

The substrate and foil lays between the two platens and the upper and lower platens are brought together under pressure over the full area being foiled (see Figure 2.4).

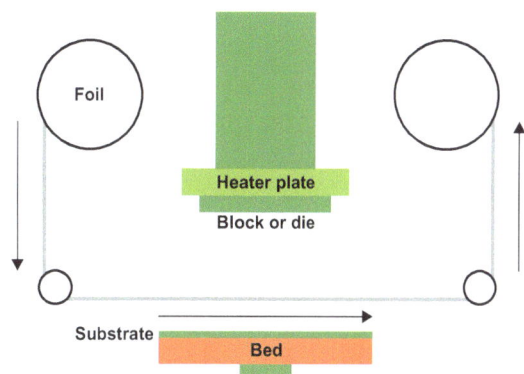

**Figure 2.4 -** Platen foiling

**Figure 2.5 -** Typical flatbed foiling/embossing unit

The platens then separate and the foil moves forward, leaving the foiled image on the surface of the substrate. A typical flatbed foiling embossing unit is shown in Figure 2.5.

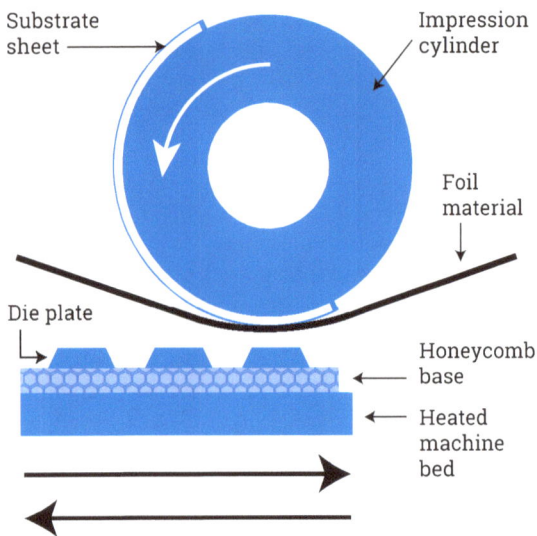

**Figure 2.6 -** Round flatbed foiling

## ROUND FLATBED FOILING

In round-flatbed foiling, the 'top' platen section is replaced by a rotating cylinder. The steel 'chase' onto which the die is attached moves forward and backwards in synchronization with the cylinder and

pressure is applied at the narrow point of pressure when the cylinder, imaged die, the foiling material and the substrate meet (see Figure 2.6).

This type of foiling system is used widely in wet glue label manufacturing when the foiling of large sheet sizes is required.

## SEMI-ROTARY FOILING

The semi-rotary foiling process is mainly used in the sheet-fed label market, where the foiling process is typically carried out on a converted letterpress machine, which is configured with a flatbed base and a large impression cylinder.

The ink duct and roller train is removed, giving excellent access to the flatbed section of the machine whilst the sheet feeder and delivery remain in position.

The foiling die mounting operation is usually carried out off-line to reduce the down time during the make ready operation. The flat foiling die is mounted on to a flatbed base and involves the use of a honeycomb base with the foiling dies secured via mounting toggles.

## ROTARY FOILING

Rotary foil stamping is one of the most popular systems of foil printing in the label industry. As this method is a full rotary system, it allows easy fitment onto rotary label presses, giving the advantage of faster running speeds and excellent foiling quality.

The biggest drawback with full rotary foiling is the cost of the 'tooling' i.e. the manufacturing and imaging of the brass foiling dies and the waste deriving from the gap which can occur between each image on the die.

There are systems however that can overcome some of the 'gapping' waste issues. These are called foil saver systems and are detailed later in this chapter.

In Figure 2.7 a typical layout of a rotary foiling unit is illustrated. The brass foiling die is positioned over a rubber impression roller and the substrate to be 'foiled' passes between these two cylinders. The foil is sandwiched between the surface of the foiling die and the substrate. A major advantage offered by the rotary system is that it does not require the same amount of impression strength as the flatbed method.

The point of contact between the foiling image, the foiling material and the substrate is narrow and therefore allows a very clean point of contact and a quick separation of the foiling material from the substrate surface, allowing very fine detail to be achieved. Figure 2.8 shows a typical rotary foiling unit mounted on a reel fed labeling press.

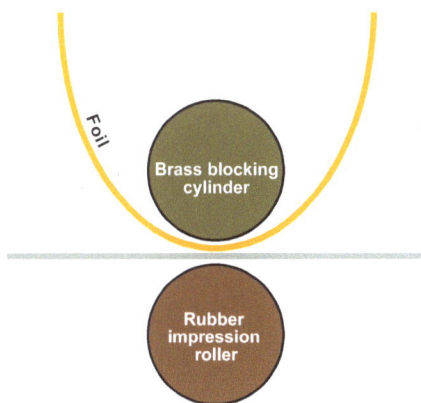

**Figure 2.7 -** Principle of rotary foiling

**Figure 2.8 -** Shows a modern hot foil unit located on a combination press

## TYPES OF DIE AND MOUNTING SYSTEMS

The manufacturing costs of foiling dies can be very high, in particular metal dies.

Metal dies (brass in particular) give a very good heat transfer and the heat can be accurately controlled, which is an important factor when producing a quality foiling result.

There are a number of different types of foiling dies used:-

- Flatbed – in magnesium, copper and brass
- Full rotary brass die
- Brass segmented rotary die – sections are used instead of a full cylinder
- Flexible metal sleeve attached to a magnetic cylinder.

## FLATBED DIES

The metal dies used in the flatbed process are not as sophisticated as those used in the full rotary system. The flatbed die does not have any curvature issues and can be etched or engraved in the flat. Once the image has been engraved or etched the die is ready for mounting in the press.

## FLATBED MAGNESIUM DIES

Magnesium is the softest of the materials used for metal dies and is the least expensive. This type of die is best suited for flatbed use, particularly when 'single' image short runs are required. (Figure 2.9)

The imaging of a magnesium plate is done using a chemical etching process. A photo sensitive coating is applied to the magnesium plate to be imaged and a film negative of the image to be produced is then placed in contact with the plate surface and exposed to a light source before being photographically developed to produce the image. The plate is then chemically etched to remove the 'non-image' area leaving the 'image' area in relief.

## FLATBED BRASS DIES

Whereas magnesium dies are chemically etched, the flatbed brass dies used in the label industry are imaged using a CNC (computer numerical control) digitally driven engraving system. This method of imaging applies to both flatbed and rotary dies.

The flatbed die is engraved in the flat. The engraving head travels over the die moving through the X and Y axis and rising and falling as the digital data instructs, to produce a very fine detailed image.

Brass is an excellent conductor of heat and is the preferred material for the hot stamping processes used in label manufacturing. The heat level of a brass die is more controllable than the magnesium and because of its durability is more suitable for longer production runs.

**Figure 2.9 -** Flatbed foiling die

### FLATBED COPPER DIES

Copper dies are imaged using an etching process similar to magnesium die etching.

The copper die is harder than magnesium and therefore more suitable for longer production runs and multiple image work. Good image etching characteristics will give excellent foiling results.

### FULL ROTARY BRASS DIE

The manufacturing of the rotary foiling die requires more engineering than the flatbed die.

The manufacturing process starts with a length of metal that is machined to the correct outside diameter of the required print length for the job to be printed and foiled (Figure 2.10).

The inner part of the die is machined out to create a tube and special end sections are fitted to the die. Oil feed connectors are fitted to these to allow heated oil to flow in and out of the center of the die. Alternatively the die can be heated using electric elements which are inserted into it and secured to the end sections of the die.

The rotary die is imaged using exactly the same

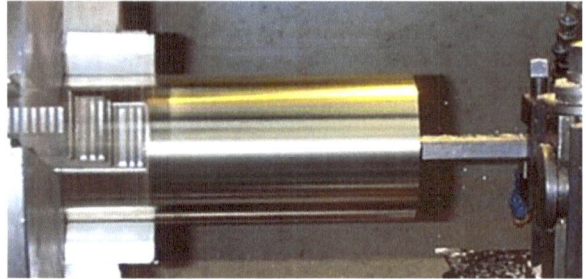

**Figure 2.10 -** Machining the die cylinder to correct diameter

principle as flatbed engraving, but instead of the engraving head traversing on the X and Y axis the engraving head moves only on the X axis. The rotary die rotates back and forth on the Y axis with the engraving head rising and falling as required. This complex system of engraving is driven by a digital file which contains the image to be engraved. Figure 2.11 shows finished rotary die with engraved images.

**Figure 2.11 -** Images engraved onto rotary die

### SLEEVED ROTARY DIES

Another option for creating a rotary foiling die involves a sleeve system. This type of die reduces the metal content of the rotary die by imaging a sleeve that is manufactured to the width of the image area required.

In Figure 2.12 the imaged brass sleeve is mounted onto a heated base cylinder before being secured. The sleeve is slid onto the cylinder and can be positioned both laterally and circumferentially to give the correct registration position The image also shows the utilization of a narrow web of foiling material even though the web width of the job being processed may be much wider.

**Figure 2.12 -** Sleeved segmented brass die (curved)

## SEGMENTED DIE

The die shown in Figure 2.13 is called a segmented die. This is a novel method of mounting individual brass dies onto a full rotary cylinder without having to use an expensive solid brass die.

The cylinder is manufactured in a honeycomb configuration which facilitates the insertion of securing 'clips' which are located onto each edge of the plate.

The curved segmented die is secured to the honeycomb base using a tensioning key to tighten the clips.

**Figure 2.13 -** Segmented die - curved brass foiling dies mounted onto a rotary honeycomb base

## MAGNETIC DIES

Magnetic foiling dies work on the same principle as the steel flexible dies used for the profile cutting of the

label shape.

The die is made up of a flexible 'shim' with a steel backing and a copper surface layer which carries the foiling image. Copper is used because of its good heat conductivity.

The imaged shim is positioned onto a heated magnetic cylinder (see Figure 2.14) and the cylinder and die are mounted in the foiling unit in the same way as a solid rotary die.

**Figure 2.14 -** Magnetic base cylinder

Imaging of this type of foiling shim is done using the same type of etching process as that used for imaging the copper flatbed die.

Fitting and removing foiling shims is a simple operation particularly as some magnetic cylinders are fitted with pins that correspond with locating holes in the shim itself, making registration an easy operation.

One of the main advantages of this type of die is the easy changeover between production jobs

## TYPES OF HEATING SYSTEMS

The use of oil or electric systems to heat the foiling die is generally one of preferential choice by the printer, as there is very little difference in the overall performance between the two systems (see Figure 2.15).

The critical factor is control of the surface temperature of the die and the evenness of the die temperature across the full width of the web.

The surface temperature of the die must be within a tolerance that is governed by the release factor of the foiling material and the optimum running speed of the job being foiled. This temperature must be maintained throughout the production run.

To establish the optimum die temperature, the printer must ensure that the pressure (impression) between the die, substrate, foil and the bottom platen (flatbed foiling) or the rubber covered impression roller (rotary foiling), is at the correct setting.

To ensure a good foiled image is achieved the foiling material must leave the surface of the substrate with a clean break and some tension needs to be maintained so that the foiling material does not become loose or 'baggy'.

At this point the printer can then adjust the impression, running speed, heat control and foil material tension to get the optimum printed result.

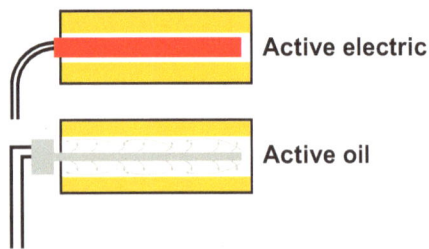

**Figure 2.15 -** Diagram showing layout for oil and electric heating of the die

## FOIL SAVER SYSTEMS

One of the downsides of hot foil stamping is the amount of wasted foiling material which occurs in the gap between each foiled image. In Figure 2.16 the foiled image area (the horse) and the gap between each image is clearly visible.

Once this reel of foiling material has passed through the press it cannot be reused.

The problem of waste foil material has been overcome by the use of 'foil saver' units. These devices use a controlled variable 'pull through' of the foiling material to minimize wastage and to give optimum use of the foil.

An incremental amount of foil is fed into a rotary foiling unit which is synchronized with press speed (die speed) through an optical encoder attached to the press drive.

The foil feed mechanism is controlled by a stepper motor which accelerates and decelerates the foil web and maintains the proper speed and tension.

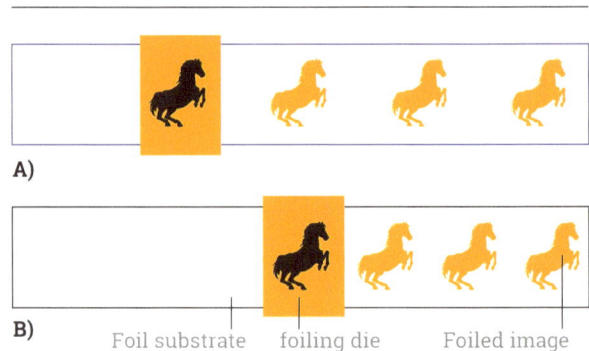

**Figure 2.16 -** The impact of using a foil saver
A) Hot foil stamping without use of foil saver system
B) Hot foil stamping with a foil saver

This movement is achieved using a method of reciprocating rollers. The roller system which carries the foil travels forward on the printing cycle and then reciprocates back, pulling the foil between the gap in each image on the die and then repeating the motion ready for each image (see Figure 2.17).

Another big advantage with this system is the facility to use individual ribbons of foil material (see Figure 2.18). This allows the printer to make optimum use of the foiling materials and make additional reductions in foil wastage. For instance the facility to utilize any narrow width 'off-cut' reels that may be held in stock.

The latest foil saver units control the movement of the foil material using a vacuum system which eliminates the need for dancer rollers. One of the big advantages with this type of foil saver system is the high speeds (20 movements per second) which can be achieved, coupled with a high degree of foiling accuracy (essential for holographic applications), when small label gaps are being printed.

## COLD FOILING

Cold foiling is an alternative method of foil stamping in which a print unit is used to apply a special adhesive

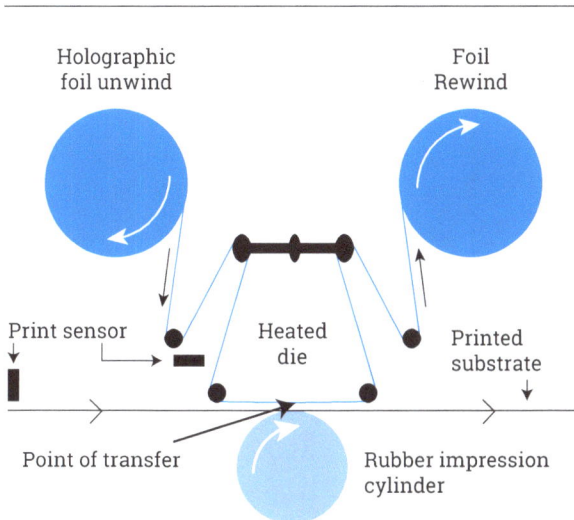

**Figure 2.17 -** Foil Saver System used for the transfer of holographic foiled images

**Figure 2.18 -** A foil saver system showing multiple webs of foiling material

on the web in the area where the metallic foil is required. When the metallic foil is brought into contact with the adhesive it adheres to it to produce the printed foil design. Figure 2.19 shows an example of cold foiling.

The use of cold foiling or 'die-less' foiling as it is sometimes known, has increased over the last decade and is now used widely in the label industry. The simplicity and low cost of adapting a label press

for the cold foiling process makes this technique easily accessible to a printer who does not want to invest in conventional hot foiling equipment. Cold foiling is a cost effective alternative to hot foiling, which requires an expensive engraved die and needs a controlled heat source which can also incur expensive running costs.

**Figure 2.19 -** Example of cold foiling

## PRINCIPLES OF PROCESS

The principle of the cold foiling process is very simple.

A standard flexo or screen print unit is used to print a UV adhesive image of the area to be foiled. This printing can be done using a standard printing plate or screen.

After the substrate has been printed with the adhesive image, it is then part-cured using UV curing, which changes the adhesive to a tacky consistency. The foiling material is positioned over the face of the substrate and the two materials are nipped together with both the substrate and the foil material running at the same web speed (see Figure 2.20).

The web then travels through a 'take off' roller and the foil is stripped away from the substrate leaving the foiled material adhered to the adhesive image leaving a perfect metallic image adhered to the substrate.

Cold foil quality depends on there being enough glue for the foil to adhere to, and absorbent substrates can be a problem. Filmic and gloss substrates are the most suitable for the cold foiling process and results comparable with hot foiling can be achieved.

The adhesive is an important part of the process and there are two different types of adhesives: free radical and cationic. The free radical adhesive reacts only when exposed to UV light whilst cationic adhesive has an initial cure to make it 'tacky' and a post cure to complete the cross linking process.

The cationic adhesive system was used extensively in the early days of cold foiling, but the system proved unpredictable and required a good press operator to get the best results. The adhesive system now favored is the free radical process because it is more user friendly and press stable.

(image) is created on the surface of the substrate to be foiled by the inkjet head, driven by a digital file which holds the image profile. No plates or dies are required using this process.

The foil material is then placed in contact with the adhesive image and the substrate and foil materials are nipped together and then exposed to a UV light source which cures the adhesive. The foil material is then stripped away leaving the foiled image (see Figure 2.21).

One of the big advantages of digital die-less foiling is the facility to produce variable data with a foiled surface finish.

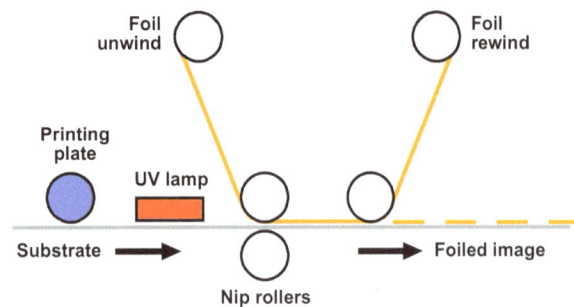

**Figure 2.20 -** The cold foiling system

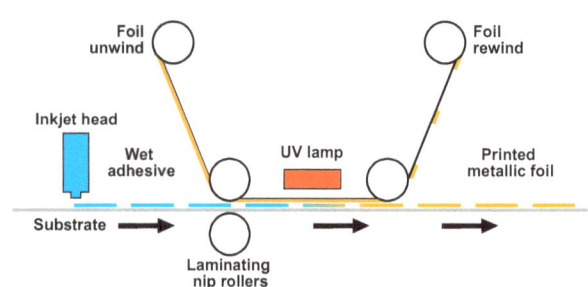

**Figure 2.21 -** Digital die-less foiling

One of the major benefits that cold foiling offers the self-adhesive label printer is the facility to run it as an 'in-line' process with the other printing processes. The cold foiling process will give faster running speeds and achieve excellent rub resistant (most important for quality specifications).

Cold foiling also allows converters to add embellishments in sectors of the label market that have previously not been viable.

One of the areas still to be fully exploited by the designer is the opportunity to foil half-tone images. This is an exciting feature of the cold foil process which uses a printing plate in the process to enable true halftone metallic effects to be created.

## DIGITAL DIE-LESS COLD FOILING

The digital die-less cold foiling process uses an inkjet system to generate the image. An adhesive pattern

## SUMMARY OF ADVANTAGES AND DISADVANTAGES OF HOT FOILING VERSUS COLD FOILING
### Hot foiling
#### Advantages
- Hot foil stamping gives a brighter more reflective color
- Hot foiling gives a wide range of quality finishes including holographic
- It is suitable for foiling and embossing combinations

#### Disadvantages
- Hot foiling is a more expensive process than cold foiling, plus the cost of dies and heating
- The prep time for hot foiling is longer than cold foiling, dies are usually outsourced
- More steps in the manufacturing chain
- Close registration issues can sometimes occur

## Cold Foiling

### Advantages

- Cold foiling units can be easily retrofitted to an existing press
- Cold foil does not require an imaged magnesium, copper, or brass die
- It only requires a simple printed image to provide the foiling area
- Cold foiling can achieve faster running speeds than hot foil stamping
- When printed as an in-line process the registration between print and the foiled area is very accurate
- The cold foiling process does not have the potential high set up costs incurred with hot foil stamping
- Halftone foiling possible

### Disadvantages

- Cold foiling does not give the very fine detailed foiled image achieved with hot foiling
- The cold foil result is not as bright and reflective as hot foiling
- Dark colored foils tend not to be recommended for cold foiling

## FOIL SPECIFICATION AND PERFORMANCE

With all types of foiling the selection of the correct foil type and the adhesive release factor is very important (Figure 2.22).

The adhesive has to be compatible with the surface of the substrate being foiled and the optimum temperature of the die surface combined with the dwell time (which is governed by the press speed) must be established and maintained throughout the job run.

The lacquer coating has to provide the correct degree of gloss finish after foiling and must meet the required quality control hardness and scuff resistant testing procedures on both rigid and flexible materials.

The optimum press speed is governed by the release factor of the foil being used. This release factor can be varied by the foil manufacturer, giving the printer the opportunity to specify the foil/adhesive release factor which will give the best results for quality and press speed.

**Figure 2.22 -** Foil specification

## FOILING - DESIGN CONSIDERATIONS

The use of metallic foils will significantly improve the added value content of the label, particularly when advantage is taken of the very wide range of face materials it can be applied to, including paper, metal, plastics and filmic.

As with all label designs which involve a printing or embellishing process, the more understanding and appreciation the designer has of the processes involved, the greater the design scope and benefits for the finished label.

It is important the label designer understands the limitations of the foiling process for both hot and cold foiled images.

Any design concept which involves a foiled image needs to consider the type size of the text and the content of any illustrations required. Fine lines and text should not be too close as this could create filling in and bridging and small type and fine serifs can cause problems. Tonal work must not be too fine and by using a coarser screen the value of the dots will be spaced sufficiently apart to allow a good foiling result.

It is possible that the foiled area may require overprinting. If this is the case consideration of the print process to be used for the overprinting needs to be carefully considered at the design stage.

## OVERPRINTING FOILS

Overprinting ink onto foils and foil substrates is a trend worthy of note. This subject is dealt with in Chapter 6 – Ink metalization.

# Chapter 3

## Holograms

The use of holograms on labels and packaging can provide a powerful visual appeal and prestige to products competing in a crowded retail space.

Eye-catching holographic materials can be used on most products, but are typically used on higher value products in the cosmetics, personal care and drinks sectors. Holograms can also perform a useful role as a brand protection or anti-counterfeiting device and are part of the global security industry's armory.

Holograms are available both as a label face-stock and as a foil that can be transferred to the surface of the label or pack, using the foiling processes covered in Chapter 2.

### WHAT IS A HOLOGRAM?

A hologram is a microscopically fine diffractive structure by which three-dimensional images are generated.

The hologram consists of two or more images which are layered in such a way that each of the images becomes visible, depending on the angle of the viewer.

The image may be made up of two layers which would comprise of a background layer and also a foreground layer. Alternatively the image may be made up of three layers, a background, middle and a foreground.

With the two-layer hologram the subject matter of the middle ground is usually superimposed over the subject matter of the background layer. This method gives a unique multi-level, multi-color effect.

Holographic label embellishments are a relatively new product to the label industry and are now widely used in labeling graphics. (Figure 3.1) The holographic

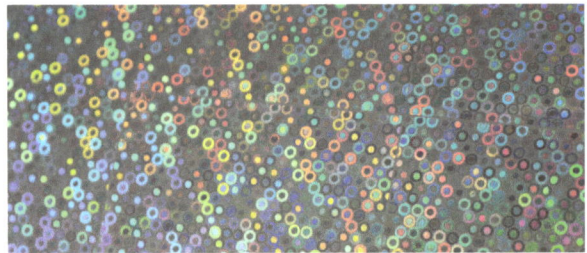

**Figure 3.1 -** Examples of holographic facestock

embellishment comprises of a 'two or three-dimensional' image which alters its position as the person viewing the image moves.

This type of imagery can produce a wide range of holographic foils which provide excellent diffraction patterns, 3D security images and anti-counterfeiting features (Figure 3.2) and because it is extremely difficult to copy a hologram, they are frequently used in high security labeling.

## MANUFACTURING THE HOLOGRAPHIC IMAGE

Holograms are two and three-dimensional photographic images that appear to have depth. The holographic image is created by superimposing a two-dimensional picture of the same subject but viewed from differing angles. This type of hologram known as a 'reflection hologram' can be viewed in normal light.

The technology used to produce a hologram can be complex and the following sections which deal with the manufacturing process have been simplified to avoid any technical overkill. Each step highlights the production of a three-dimensional holographic image.

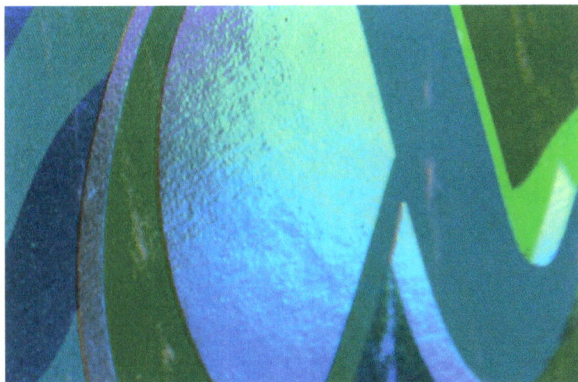

**Figure 3.2 -** Examples of holographic materials used as decoration in labeling and packaging

## IMAGING

The following section explains in simple terms how a holographic image or pattern is created.

The subject of the holographic image is illuminated using a laser beam, allowing reflected light from the laser to fall on a photo-resist plate. At the same time a reference beam from the same laser is directed onto the photo-resist plate. These two light beams react with the photo-sensitive coating and record a holographic image of the subject (see Figure 3.3).

It is most important that there is no movement of the subject during this process as this will produce an out of focus image.

The imaged photo-resist plate now holds the 'master' holographic image and is then processed using a photographic developing solution. After developing the surface of the plate it is similar to the surface of a gramophone record with the image made up of very fine grooves.

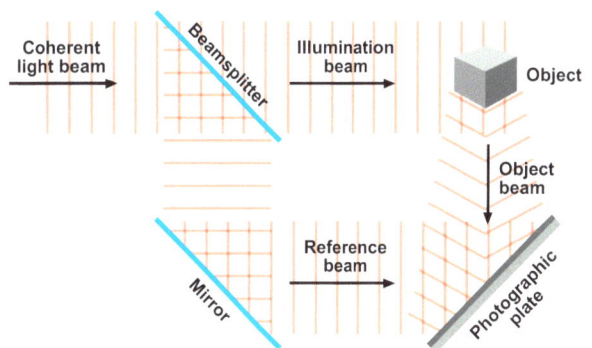

**Figure 3.3 -** Illuminating and transferring the image

## ELECTROPLATING

The next stage in hologram production is the electroplating process. This gives the master plate a much more robust surface. To produce a good electrical conductivity, the master is sprayed with silver paint and then immersed in a nickel tank.

An electric current is passed through the tank and a thin nickel coating is applied to the master plate.

The plate is then washed and the thin nickel coating is removed from the master plate. This nickel shim is called the 'master-shim' and holds a 'negative' image of the original subject.

By applying the same process of nickel coating using the 'master-shim' a 'positive' imaged shim can be produced. This in turn is used to produce 'negative' shims which are then converted to stamper shims that are used to print/emboss the holograms.

## THE EMBOSSING PROCESS

Polyester film with an acrylic coating is used as the vehicle for the holographic image. The 'master-shims' are located into an embossing unit and by applying high pressure combined with heat, the image from the shim produces the hologram in the polyester film.

## METALIZING AND FINISHING

Metalization refers to the coating of the embossed polyester film surface with a highly reflective metal coating.

The embossed roll of film is placed in a vacuum chamber and the air is removed. An aluminum wire is positioned in the chamber and the wire is heated to 1100°C which vaporizes the aluminum wire, which in turn coats the surface of the polyester with aluminum particles. The coated film is removed from the vacuum chamber and a coating of lacquer is applied to the film that allows the holographic image or pattern to be overprinted if required. The holographic film is applied to a self-adhesive substrate using a hot foiling method of application, or alternatively a pressure-sensitive adhesive is applied to the back of holograms which in turn can be applied to a self-adhesive label.

## APPLYING THE HOLOGRAM

The application of a holographic image onto the top surface of a self-adhesive substrate is done using a foil-saver type system (see Chapter 2 for more information).

This operation can be carried out as an off-line or on-line application using either a rotary or flatbed hot foiling method.

To ensure accurate registration of the hologram with the printed design on the label, the foil that carries the holographic image must be synchronized with the printed web of the self-adhesive substrate.

Sensors monitor the position of the holographic image prior to entering the foiling head and the foil-saver unit will slow down or increase the speed of the holographic web to ensure that the two webs are exactly synchronized (see Figures 3.4 and 3.5).

## MARKETS

Holograms and holographic face-stocks are widely used to introduce light refractive effects and iridescence to labels and packaging. As well as adding a stunning decorative feature to high-end consumer products, holograms also have potent anti-counterfeiting and brand protection properties.

They offer a wide variety of different features which can be matched to a range of security requirements,

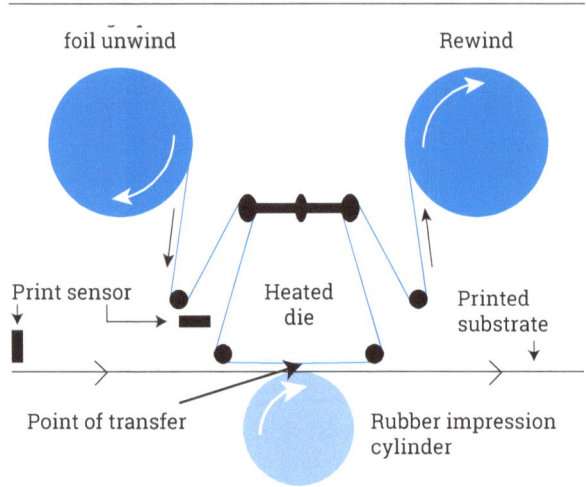

**Figure 3.4-** Hologram application via a foil saver

**Figure 3.5 -** In-line foil saving unit featuring registered hologram patch insetting.

from those used in relatively low-cost commercial applications such as packaging and labels, to a more sophisticated security method of protecting currency.

Holograms incorporating both overt and covert

machine readable features, variable data and unique serial numbering have become powerful tools in the prevention of counterfeiting of documents, labels, seals and tags.

In summary, holograms used for anti-counterfeiting and brand protection have a number of benefits;

- Virtually impossible to replicate
- Cannot be reproduced by photocopiers or scanners
- Will expose any attempts at alteration or removal from a document or label
- Are easy to authenticate with the naked eye
- Require specialist knowledge and equipment to originate and manufacture
- Cannot be replicated by reproduction using any conventional printing process

# Chapter 4

# Embossing

Embossing is a technique that adds a distinctive elegance and a tactile quality to any paper based label-stock or substrate.

It is used to raise a specific part of the label design above the label surface to create a new dimension to the pack design.

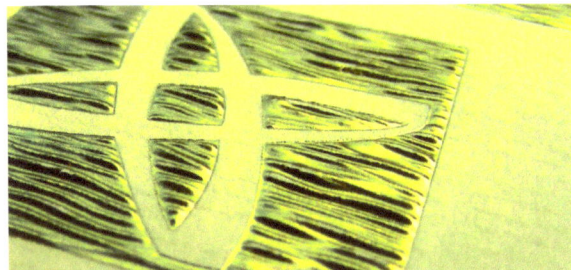

**Figure 4.1 -** Examples of embossed images

## THE EMBOSSING PROCESS

Embossing is one of the most commonly used embellishments in the label industry. This is the process of creating a raised relief 3D image (see Fig 4.1) or alternatively a sunken image (de-bossing) on the label substrate.

The embossed image is created using a male and female die. The female die has an engraved or etched recessed image whilst the male die has a raised image.

The substrate being embossed is sandwiched between the male and female dies and pressure is applied which forces the male relief image into the female recessed image. This action pushes the label substrate into the recessed female image creating a raised profile on the label surface (see Fig 4.2). Heat may also be applied to assist in producing the optimum embossed image.

No ink is used for the actual embossing process but very often the embossed image will lay over the printed area of the label. This will involve close register between the printed image and the embossed image and this is called a 'registered emboss'. If the embossed image lies in a non-printed area this is called blind embossing'.

## DEBOSSING

Another type of embossing process is 'de-bossing' which uses exactly the same method as the

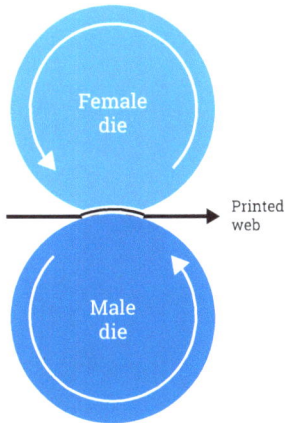

**Figure 4.2 -** The embossing process

embossing process, but the function of the male and female dies are reversed.

Whereas an embossed image is created with a 'raised' image in the label substrate, de-bossing is a 'recessed image' in the surface of the substrate. The process involves the application of pressure to the face side of the label substrate, forcing the material downwards into the female thereby creating the recessed profile (see Figure 4.3).

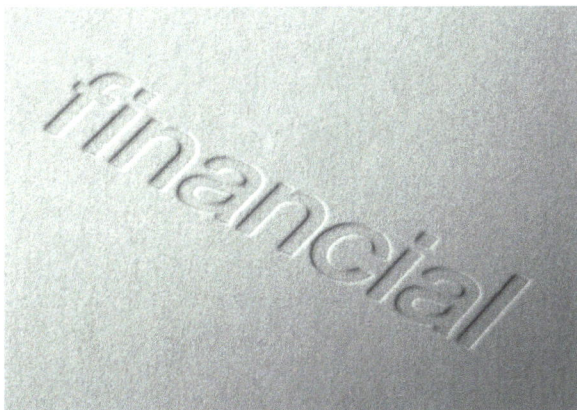

**Figure 4.3 -** Example of de-bossed image

## METHODS OF EMBOSSING

The embossing process can be undertaken using a flatbed, semi-rotary or a fully rotary system (see Figures 4.4a, b and c). The choice of method is generally governed by the type of label required, the run length and the cost of the dies.

Flatbed

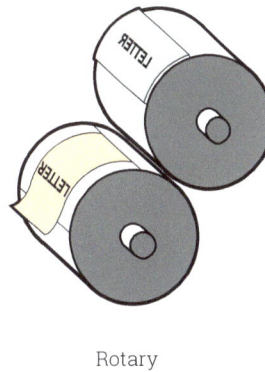

Rotary

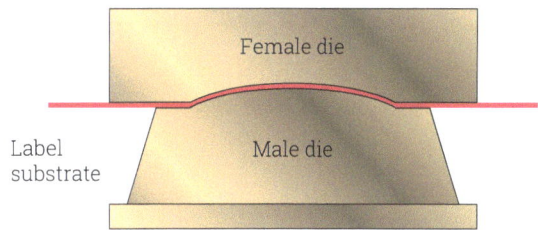

**Figure 4.4 -** Embossing systems

## FLATBED EMBOSSING

The flatbed embossing unit works on the same principle as the flatbed die-cutting unit and the flatbed foiling unit.

The illustration above (Fig 4.4a) shows the flatbed type of press configuration. The substrate, which can be in sheet-fed or web fed format is positioned between the two plates which hold the embossing dies. The top plate moves up and down sandwiching the substrate to be embossed (as indicated by the arrow).

Because of the heavy construction and high impression strength of this type of 'platen' configuration, it is used extensively for both embossing and die-cutting in both sheet-fed and reel fed label manufacture.

Figure 4.5 shows a typical flatbed embossing unit with the top and bottom plates holding the embossing dies visible.

**Figure 4.5 -** Flatbed embossing unit

## SEMI-ROTARY EMBOSSING

The semi-rotary embossing process is mainly used in the sheet-fed label market. The embossing process is usually carried out on a converted letterpress machine, which is configured with a flatbed base and a large impression cylinder. The ink duct and roller train is removed giving excellent access to the flatbed section of the machine, whilst the sheet feeder and delivery remain in position.

In Fig 4.6 the flatbed base and the large

**Figure 4.6 -** Semi-rotary embossing unit

impression cylinder are clearly visible with the sheet feeder positioned at the top. The female flat die is mounted onto the flatbed base. This usually involves the use of a honeycomb base and the dies are secured with mounting toggles (Fig 4.7).

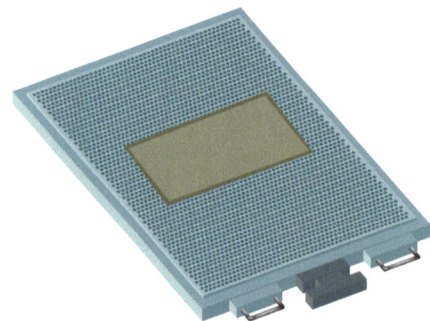

**Figure 4.7 -** Honeycomb base with embossing plate mounted

The mounting operation is usually carried out off-line to reduce the down time during the make-ready operation.

The interesting thing about this type of embossing is that the male element of the process is created by the press operator, involving the following process sequence;

1. The dies are mounted onto the honeycomb

base and positioned in the press
2. The dies are manually inked
3. The cylinder is dressed with card or blotting paper
4. The operator feeds a sheet through the press with the cylinder on impression allowing a print to be taken of the die showing if the dies are in register. If any of the dies are not in register then the operator re-positions them, using the toggles to make any adjustment.
5. The process is repeated until all the dies are in correct register
6. The card and blotting paper are removed and a special embossing make ready material is fixed to the cylinder. This material is dampened and the press is then turned over under impression
7. The material is forced, under pressure, into the female die and the press is run for several minutes until the male image is formed and the material has dried and hardened. This process has created the male embossing element of the process
8. If any individual areas require additional pressure the operator can paste tissue onto the specific area, thereby ensuring that the substrate being embossed is fully contacting the base of the female die

## EMBOSSING ON INTERMITTENT FED PRESSES

With intermittent feed presses a flatbed embossing unit is used. The embossing head is positioned to allow the web to travel between the two embossing dies. The web travels with a stop-start motion and the embossing takes place when the web is stationary.

As the web stops, the embossing dies come together to create the embossed image. The plates then open and the web moves forward. This action is called the 'pull' and the distance of the pull is set by the printing width of the plate cylinder (see Figure 4.8).

## FULL ROTARY EMBOSSING

Rotary embossing is one of the most popular systems in use in the label industry, because it allows easy fitment onto rotary label presses, thereby giving the

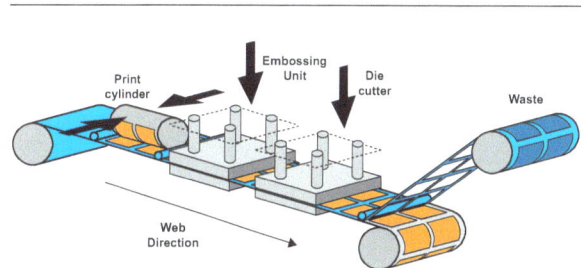

**Figure 4.8** - Embossing on intermittent fed press

**Figure 4.8a** - Rotary embossing unit on a modern label press

advantage of faster running speeds and excellent embossing quality (see Fig 4.8a).

The biggest drawback with full rotary embossing is the cost of the 'tooling' i.e. the manufacturing and imaging of the embossing dies. The configuration of rotary embossing on a rotary label press (Figure 4.9)

There are four factors that control the quality and consistency of good embossing.

- Die depth
- The applied pressure
- Type of substrate
- The use of heat

Each of these factors is explained in more detail in the following section.

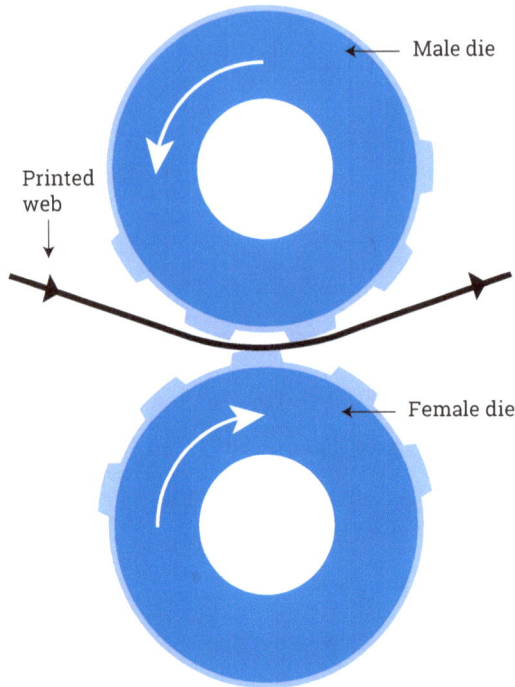

**Figure 4.9 -** Full rotary embossing

### 1. The depth of the female die

In the 'embossing' process the depth of the female die will govern the height of the embossed image - the deeper the die the greater the profile of the image. Consideration must be given to the type of substrate to be embossed and what depth can be achieved before the substrate fibers break. The required depth can be assessed using the original artwork allowing the engraver to determine the correct die depth.

### 2. Applied pressure

Quality embossing requires considerable pressure to be applied to the dies during the embossing process. It is important that the substrate contacts the bottom of the female die and this can only be done effectively by pressure on the male die.

If there is insufficient pressure and the dies do not fully contact then some of the embossed image,

particularly in the fine detail may be lost. Too much pressure will damage the surface of the substrate and can in some cases puncture the substrate, destroying the embossed area.

The two methods of embossing, flatbed and rotary, can vary in the amount of direct pressure required. Flatbed embossing needs far greater 'overall' pressure (tonnage) than rotary.

The amount of pressure needed can be reduced by using a 'rocking' technique. The top plate, which

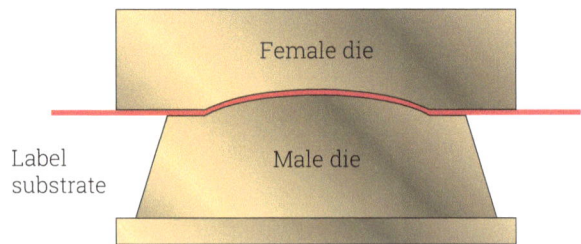

**Figure 4.10 -** Principle of applying pressure on an embossing unit

holds the male or female die (dependent on whether it is a de-bossed or an embossed image) is set to contact the bottom plate using a rocking motion. This allows the two embossing dies to make contact in stages, thereby reducing the amount of pressure required to achieve the correct embossed depth.

Rotary embossing automatically benefits from this principle and therefore requires less pressure than the flatbed process, which requires heavy pressure.

### 3. Substrate and heat

The application of heat can be an advantage during the embossing process. The use of a heated die can aid the molding of the substrate fibers during the embossing process, but care must be taken not to apply too much heat, as this will distort the substrate.

The type of substrate to be embossed is also an important consideration for the label designer and converter. Substrates vary in the amount of pressure they can absorb before the material fibers break and the substrate surface is damaged.

The function of the embossing die is to mould the paper fibers into the profile of the embossed image without damaging the substrate surface, but allowing the embossed image to maintain its profile after the process is completed.

Synthetic materials are unsuitable for embossing as they are cannot be used to create a raised image and these substrates will not accept any direct pressure from an imaged die and are easily punctured.

## TOOLING AND ENGRAVING

There are **five** different types of embossing dies used in the label industry.

The flatbed and semi-rotary processes use steel, magnesium, copper and brass dies, whereas full rotary embossing generally use only brass or steel dies.

The manufacturing costs of embossing dies can be high, particular when tooling and engraving full rotary male and female brass dies.

## FLATBED EMBOSSING DIES

The metal dies used in flatbed embossing are not as sophisticated as those used in the full rotary system. The flatbed die does not have any curvature issues and can be etched or engraved in the flat. Once the image has been engraved or etched the die is ready for mounting in the press.

The characteristics of each of the die materials are reviewed below.

### • Flatbed magnesium embossing dies

Magnesium is the softest of the materials used for metal dies and is the least expensive. This type of die is best suited for flatbed use, particularly when 'single' image short runs are required.

The imaging of a magnesium plate is done using a chemical etching process. A photo sensitive coating is applied to the magnesium plate to be imaged. A film negative of the image to be produced is then placed in contact with the plate surface and exposed to a light source, before being photographically developed to produce the image. The plate is then chemically etched to remove the 'non-image' area leaving the 'image' area in relief.

### • Flatbed brass or steel dies

Unlike magnesium embossing dies that are chemically etched, flatbed brass and steel embossing dies are imaged using a CNC digitally driven engraving system. This method applies also to the imaging of rotary dies made from brass or steel.

The flatbed embossing die is engraved in the flat and the engraving head travels over the die moving through the X and Y axis and rising and falling as the digital data instructs. This method of engraving will produce a very fine and detailed image.

### • Flatbed copper dies

Copper embossing dies are imaged using the etching process similar to magnesium die etching. Due to the fact that the copper embossing die is harder than magnesium it is more suitable for longer production runs.

The copper die is also better suited for multiple image work and with good image etching characteristics it will give an excellent embossed result.

## ROTARY EMBOSSING DIES
### • Rotary brass or steel dies

The manufacturing of the rotary embossing die requires more engineering than the flatbed embossing die. The manufacturing process starts with a length of brass that is machined to the outside diameter of the required print length for the job to be printed and embossed (see Fig 4.11 ).

The ends of the cylinder are then machined to create the end assemblies, the dimensions of which are dependent on the type of press and the specification of the embossing unit being used.

The embossed image (both male and female) is imaged using exactly the same principle as flatbed engraving, but instead of the engraving head traversing on the X and Y axis, the engraving head moves only on the X axis.

The rotary die rotates back and forth on the Y axis with the engraving head rising and falling as required. This complex system of engraving is driven by a digital file which contains the image to be engraved

Figures 4.12 and 4.13 show examples of final imaged embossing cylinders with close-up of positive and negative images.

**Figure 4.11 -** Pair of imaged embossing cylinders – positive and negative

### • Sleeved rotary embossing dies

Another option for rotary embossing involves what is known as a sleeve system.

This system creates a sleeve which slides over a reusable base cylinder, thereby reducing the volume of metal required to produce the embossing die. The sleeve is manufactured to the width of the image area required.

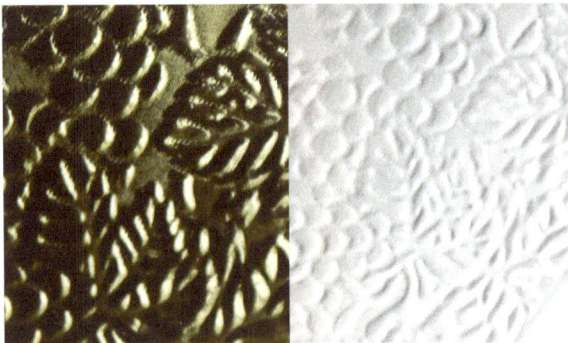

**Figure 4.12 -** Examples of combination foil/embossing

Once the sleeve has been slid onto the base cylinder it can be positioned both laterally and circumferentially to give the correct registration position.

**Figure 4.13 -** Close-up of imaged embossing cylinders

## COMBINATION FOIL-EMBOSSING

The embossing process can also be combined with hot foil stamping and this is called combination foil/embossing. The process involves embossing and foiling in one operation using a single combination brass die. The die has a sharp cutting area around its edge to ensure that the foil substrate strips evenly on each embossing cycle.

Combinations of foiling and embossing are still carried out in label decoration, but they were used extensively for the manufacture of 'seals' for both self-adhesive and non self-adhesive applications

Seals provided a superb and unique method of product decoration, particularly on high weight metalized substrates.

Fig 4.14 shows an example of a substrate which has been blind embossed (right) and also combination foil embossed (left).

## COMBINATION PRINTING AND EMBOSSING

Die stamping is a method of printing and embossing the image using the same engraved die. The die is manufactured in steel or copper. The flatbed male and female dies are mounted in the same way as normal embossing dies and placed in a die-stamping press. This press usually has a letterpress roller inking system which deposits a film of ink onto the surface of the female die. The substrate to be die stamped is positioned between the two dies which are then pressed together under extreme pressure leaving a printed and embossed raised image.

## HOLOGRAPHIC EMBOSSING

Holographic embossing is used extensively for the mass production of hologram images.

The process reproduces very fine holographic patterns onto a polymer substrate carrier which provides the reflective metallic effect and brightness of the hologram.

The process for producing the embossed hologram involves passing the reflective polymer substrate between the die pressure roller onto which is fixed the holographic nickel shim. A pressure roll makes contact with the shim and with a high temperature and pressure the image is transferred onto the surface of the polymer and thus producing the holographic image. For more information on holograms and their production see Chapter 3.

# Chapter 5

# The bronzing process

The bronzing process is a method of embellishing the surface of a label with a distinctive metallic, burnished image (usually gold).

Bronzing is a relatively slow and expensive process used predominantly for wet glue label customers in the high quality food, wine spirits and beverage segments. The bronzing process is also used extensively on greeting cards and for tobacco packaging.

Used in Europe and the USA, today the growth markets for bronzing are in Asia, China, India and Indonesia, where it is also used for security purposes and as an anti-counterfeiting measure.

## THE PROCESS
The label or sheet of labels is printed (usually by letterpress or litho) with a special adhesive ink known as 'prep'. This bronze 'prep' (preparation) is a varnish which contains a colored pigment. Whilst the image is still wet a metallic bronze powder is applied to the surface. The 'prep' dries by the oxidation process thus sealing in the bronze powder.

The excess powder which has not adhered to the image is then vacuumed off the substrate and re-cycled back into the bronze powder reservoir. A final dust removal is then carried out to ensure that no residual powder is left on the substrate surface.

The sheet or web is then burnished using rotary burnishing bands of soft material which polishes the image giving the appearance of gold.

A further pass through the press may then be required to apply a varnish coating to the bronzed area to ensure the correct scuff resistance is met to meet agreed quality standards.

The bronzing process is illustrated in Figure 5.1 below.

## EARLY BRONZING
In the early days of bronzing the process was carried out by hand and was generally for short run small sheet work printed on hand-fed letterpress platen presses.

A wad of cotton wool was used for dusting the bronzing onto the printed image. The powder was lightly dusted on to ensure there was no smearing of the printed image and the surplus powder was the tapped of the sheet which was left to dry, allowing the bronze powder to be fully secured to the printed image

When the bronzing 'prep' is dry, the loose bronze

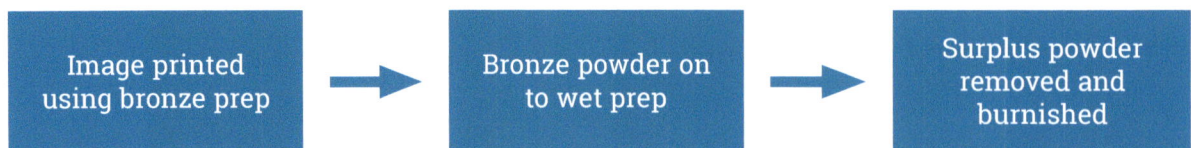

| Image printed using bronze prep | → | Bronze powder on to wet prep | → | Surplus powder removed and burnished |

**Figure 5.1 -** The bronzing process

was fully removed and the bronzed image is then burnished to give the unique gold finish. Figure 5.2 and 5.3 show examples of bronzing to produce a metallic effect on spirit labels.

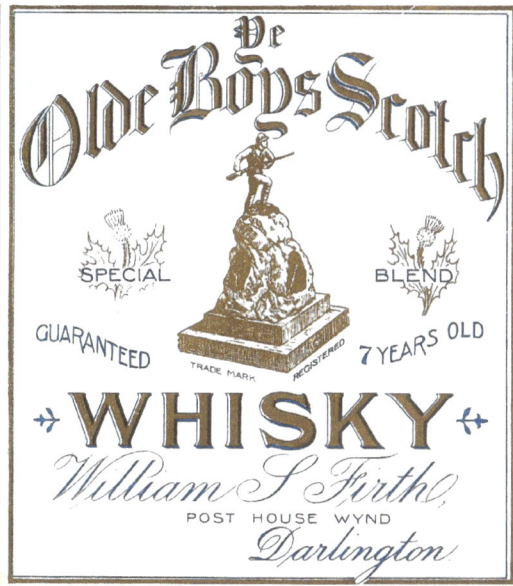

**Figure 5.2 -** Example of early bronze label

**Figure 5.3 -** Example of bronzing

## THE MODERN BRONZING PROCESS

The modern automated bronzing machine is a stand-alone unit which houses the bronze powder, the application system, the powder extraction and filtering and the burnishing belts.

The unit can be easily moved to operate in tandem with the selected choice of printing process and the type of press to be used for printing the 'prep'.

The flatbed type of bronzing machine operates with the sheet in the flat position during the bronzing process.

**Figure 5.4 -** Modern bronzing units used for sheet-fed labels.

The sheets pass over the rubber blanket through the bronzer and the flow of bronze powder onto the tacky image is controlled through a duct similar to a conventional printing press ink duct.

A series of rotating dusting bands pass over the dusted sheet to remove the surplus bronze and burnish the image area. These dusting bands make contact with brushes which are positioned at each end section allowing the dusting bands to be brushed clean as they rotate.

A final dusting and burnishing is carried out prior to the sheets exiting the machine. This is done by rotating rollers covered with a soft material.

These machines feature a powerful internal vacuum system which draws the surplus bronze from the dusting section of the machine and the powder is filtered and collected to be used again (Figure 5.3).

## BRONZE POWDER

Bronze powder is made from brass, zinc alloy and copper platelets. Molten metal is processed to produce tiny platelets and the platelets are graded for size, greased and polished.

The use of different metals to produce the platelets gives differing tones of bronze powder, plus copper based powders and silver aluminium. Other colors can also be produced by dyeing the platelets with colored dyes.

It is important that the bronze powder is fully secured to the printed image. Absorbent papers can be a problem as the prep can be absorbed into the substrate leaving insufficient prep on the surface of the substrate to hold and secure the bronze powder.

If the powder has not fully adhered to the prep the bronze will rub off and the image will be smeared. This damage to the printed image will occur when the sheet/web is burnished. The problem of loose bronze is also a serious problem when the label is used for food packaging.  Any loose bronze powder would mean that the job would be rejected.

To help overcome the problem of loose bronze it is quite common for the bronzed image to be over-varnished. This seals the bronze powder and eliminates any loose bronze particles.

## BRONZING DEVELOPMENTS

Although the bronzing process gives a 'unique' metallic finish it has always been considered to be a 'messy' process. As the bronzing powder is made up of very fine particles, airborne contamination can occur, which can lead to powder being distributed around the area of the bronzing operation and is a potential health risk. This situation has led to the development of extraction systems which are now used on modern bronzing equipment, which gives very efficient dust extraction and control.

The efficiency of the modern bronze extraction system has been confirmed by studies which have favorably compared the amount of waste generated during a bronzing operation to that from the hot foil stamping process. Worthy of note is the fact that waste bronze powder, which consists of 85% copper, has a resale value and can be sold to the recycling industry.

## PRINTED BRONZE

The issue of loose powder has prompted the label industry to seek out alternatives to the bronzing process. In recent years, development work has been carried out which allows the replacement of the traditional bronzing process with a conventional printing processes.

This development involves immersing a standard bronze powder into a conventional 'solvented' clear varnish ink system and then printing the bronze ink using a conventional screen process. This method eliminates the requirement for a secondary varnish pass to seal the bronzed image whilst overcoming some of the environmental/health and safety issues surrounding the use of loose bronze powder.

## FOILED BRONZE

Developments using bronze effect foil applied using hot or cold foiling have produced acceptable results, but these systems do not fully achieve the unique effect created by the traditional bronzing process.

## WEB-FED BRONZING

A technical solution developed by Matheoschat allows a web-fed bronzer to be integrated into a self-adhesive label printing press. This unit is capable of delivering specially developed gold and silver prep onto a variety of substrates.

New powders are also available which give new shades of gold, and silver plus pearlescent and other special effects.

**SHEET-FED BRONZING**

Within the sheet-fed market there is a demand for small bronzing machines that cater for the small format, short run markets.

Mobile lightweight bronzing units have been developed that can be connected and disconnected to a printing press quickly. After the bronzing work is completed the unit can then be removed from the press.

# Chapter 6

# Ink metalization

Gold and silver foil in particular have long been used on packaging for products such as spirits, cosmetics and personal care where their high retail value has been able to withstand the additional cost of what was a relatively expensive process.

In recent times however, developments in ink systems that seek to mimic metalized substrates, foil stamping, bronzing and foil transfer, are reducing costs and therefore widening the opportunities to add stunning surface embellishments to a wider range of shorter run and lower value products.

This chapter will explore the subject of ink metalization and the new options that are now available.

## HISTORY - THE INTRODUCTION OF METALLIC INKS

Evolving from the production of gold bronze and aluminum metallic powders for bronzing, Eckart, along with other companies, moved into the development and then production of cutting edge metallic inks with manufacturing facilities in Europe and North America.

These new generations of metallic inks were then used in the production of high-class labels, wrappers, box tops, covers, greeting cards and other work where bright, high impact effects were appropriate.

Unlike bronzing where a base size color was first printed and then dusted with a metallic powder, the metallic powders were incorporated into a liquid carrying medium (vehicle) to provide a press-ready metallic ink. Some of the early metallic ink firms

supplied the powder and vehicle separately for mixing in small quantities fresh at the press, with improved results.

Metallic ink printing was more challenging for the press operator to control than conventional ink. One reason for this was that the metallic powder blended into the ink mixture and could not be ground as fine as other pigments, because the metallic ink would lose its lustre. The larger particles created problems on the press, especially with the offset process.

## METALLIC INKS

Metallic inks consist of powder-like metallic flakes, such as aluminum and copper alloys, which are mixed with a suitable varnish or pigment carrier. The varnish or carrier dries rapidly and binds the flakes to the surface.

Coated papers give the best results, while gravure, due to the ink film thickness, produces the best optical effects with gold and silver metallic inks. On rough surface papers, a base ink is usually printed first, allowed to dry and then overprinted with gold.

Metallic gold and aluminum inks can be printed by the offset, gravure or letterpress printing processes, although care should be taken to use a neutral fountain solution when printing offset to avoid tarnishing (see Figure 6.1).

**Figure 6.1 -** Metallic effects can be replicated by using inks.

## VACUUM METALIZED PIGMENTS

As mentioned, metallic inks contain metallic flakes. Ideally these should be flat platelets which when printed, all lie down flat to assemble a mosaic-like mirror. In reality however, the particles are not platelets and they don't all lie flat and so most metallic inks only give a very dull imitation of the effect they try to mimic.

The development of inks based on vacuum metalized pigments (VMP) exhibiting mirror finishes, effectively solves this problem (Figure 6.2).

**Figure 6.2 -** Vacuum metalized pigments (VMP) exhibiting mirror finishes

## VMP PIGMENTS AND THEIR PRODUCTION

VMP pigments are produced by vacuum evaporation (physical vapor deposition) of a microscopically thin aluminum film onto a carrier foil. The aluminum layer is subsequently removed to be broken into particles for incorporation into various ink media. This process makes it possible to produce very thin and light pigments with a homogenous particle thickness hundreds of times less than that of conventional metal pigments. Because they are lighter and thinner these pigments can align themselves parallel to the substrate surface faster than conventional metal pigments. When formulated correctly, these properties result in the formation of a smooth and even surface with very few edges and corners to scatter light. Figure 6.3 clearly shows the improved structure and alignment that can be achieved with a VMP flake versus a conventional pigment structure.

### CONVENTIONAL ALUMINIUM PIGMENT

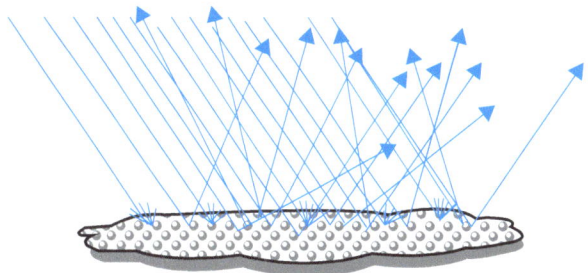

**Electron-micrograph of conventional aluminium flake**

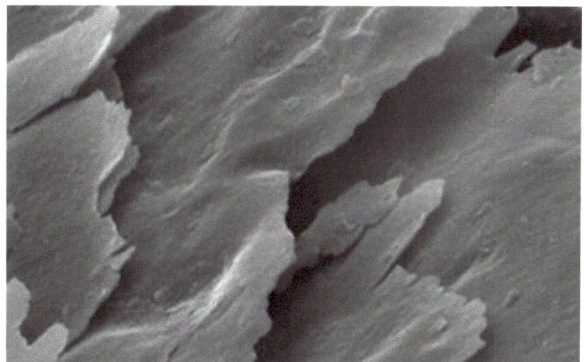

## VMP FLAKE

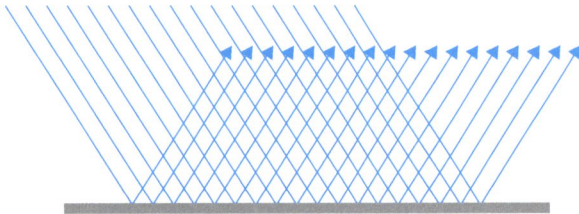

### Electron-micrograph of VMP flake

**Figure 6.3 -** The improved structure and alignment that can be achieved with a VMP flake versus a conventional pigment structure.

The steps involved in the VMP manufacturing process are illustrated in Figure 6.4.

## PRINTING VMP METALIZED INKS

Press ready VMP metallic inks can be applied by all the major printing processes and are available in oil, solvent or water-based formats.

Whilst there are water passivated grades of VMP, which are specially treated to ensure stability of the aluminum for use in water-based coatings, solvent based systems tend to generate the best results. The benefits of using VMP inks versus other types of metallic decorating technology are mostly in cost savings and ease of use.

### STEP 1: RELEASE COAT

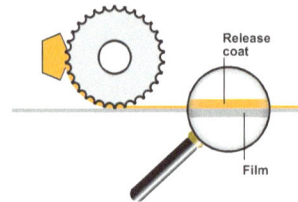

A release coating is applied to a carrier substrate e.g. polyester

### STEP 2: PVD

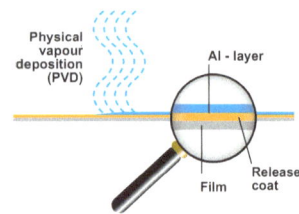

Aluminum is then applied to the carrier substrate by physical vapor deposition (PVD)

### STEP 3: STRIPPING

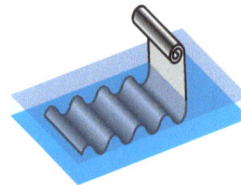

The metallized carrier substrate is then passed through a solvent bath. The release coat dissolves & the metallization breaks off into the bath to be collected

### STEP 4: PARTICLE SIZING

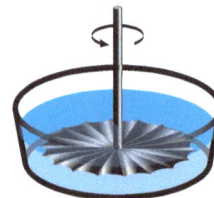

This step involves breaking the metallization layer into particles of similar geometry

**Figure 6.4 -** VMP Manufacturing Process

Improved production speeds associated with printed metallics form an important part of the relative cost equation.

VMP inks are likely to be cost effective versus other types of decorating when they are printed selectively, but this varies by application and depends on many factors.

VMP is used extensively for wet glue applied beer label applications where gravure printing is commonly employed. Metallic inks are understood to have positive benefits when it comes to pack recyclability. The discreet ink particles make for easier penetration of the caustic solution and therefore allow for easy stripping of the label from the glass bottles.

Inks containing VMP's are also experiencing growth for 'no-label look' self-adhesive applications. Reverse printed gravure onto a clear substrate such as polypropylene, means that VMPs are capable of achieving a bright and highly reflective appearance that resembles that of foil blocking.

The thick ink coverage possible with the rotary screen printing process enables the larger particle sizes associated with metallic inks to be more easily applied to the surface of the label or substrate.

As with the gravure printing process, solvented screen printing is able to float VMP pigments on a layer of solvent that is then dissipated by evaporation. The flakes align better to a clear substrate when reverse printed, thereby producing a high degree of reflectivity and brilliance, which is further enhanced (and protected) by the natural gloss of the filmic material.

The printing of metallics using both the rotary screen and gravure printing processes are illustrated in Figure 6.5.

Both gravure and solvented screen systems require investment in extraction and a controlled explosion-proof environment, but for short runs screen printing can offer a number of advantages over its gravure counterpart.

A rotary screen printing system allows the re-use and re-imaging of screens. The ability to re-use a screen up to 15 times significantly reduces the costs per job, since the initial screen cost is apportioned to the number of jobs it is used for.

**Rotary Screen Printing**

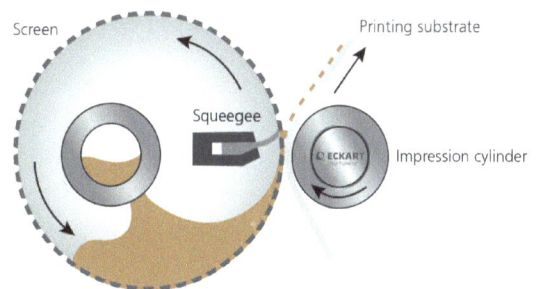

**Gravure Printing**

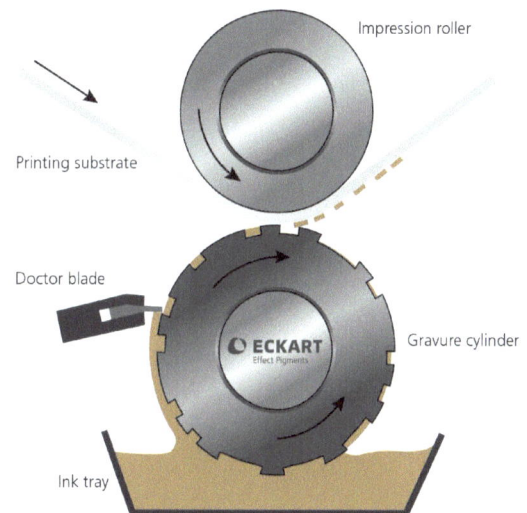

**Figure 6.5 -** Printing VMP Rotary Screen versus Gravure

For longer runs and where screen wear can be an issue or for achieving subtle vignettes and halftones, the gravure cylinder can be a better solution.

The flexibility of screen modules can also be advantageous. The rotary screen head typically will slot into fixed positions on most printing presses and therefore can be used to deliver a wider range of ink systems and coating weights, which can be important in achieving a desirable metallic design feature (see Figure 6.6).

**Figure 6.6** - Examples of VMP rotary screen printed metallics

## ROTARY SCREEN VMP - SUMMARY OF BENEFITS

- Solvented screen floats VMP particles on layer of solvent
- Solvent offers optimum alignment of VMP platelets
- Screen can be re-used up to 15 times
- Flexible system for narrow web printing

## MORE VMP METALLIC DESIGN OPTIONS

As designers become more familiar with the capabilities of VMP inks, more design options are emerging such as printing screens and vignettes and trapping colors over silver to create new effects.

Metallic inks for digital printing are now also available. These solvent-based inks are suitable for many standard piezo inkjet print heads.

Low migration metallic pigments, specially designed to comply with the strict EU legislation for food packaging, are also available.

## PRISMATIC EFFECTS

New ink systems are emerging that are able to create a holographic or prismatic effect using a printing plate. The addition of micro embossing to the VMP (vacuum metalized pigment) manufacturing process makes it possible to create a rainbow-effect metallic ink (Figure 6.7). These special refractive pigment particles have been formulated into solvent and UV base inks for most printing processes.

Typically available as UV or solvent-based flexo inks, these ink systems eliminate the need for costly holographic substrates. The technology uses the flexographic process in combination with photopolymer printing plates to transfer the desired image to the substrate. UV curable ink or adhesive is used, with the holographic image laminated over the wet uncured ink or adhesive. The structure passes through a UV lamp and the ink or adhesive cures through and bonds the holographic material to the base substrate.

Applications include any that require foiling or the appearance of a holographic effect, such as cosmetics, drinks labeling and antifraud use.

**Figure 6.7** - Example of prismatic effect inks

## METALLIC DOMING VIA ROTARY SCREEN (GALLUS SCREENY)

Metallic doming is a reflective, metallic raised structure manufactured using a relief in rotary screen printing (Figure 6.8).

The lasting metal-relief effect can be achieved on paper or plastic film, such as tube laminates or transparent self-adhesive material.

The metallic doming method uses rotary printing to apply the printed image. A metallic foil is then laminated over the relief. The process allows printed images with metallic reliefs (embossed effects) to be produced on plastic.

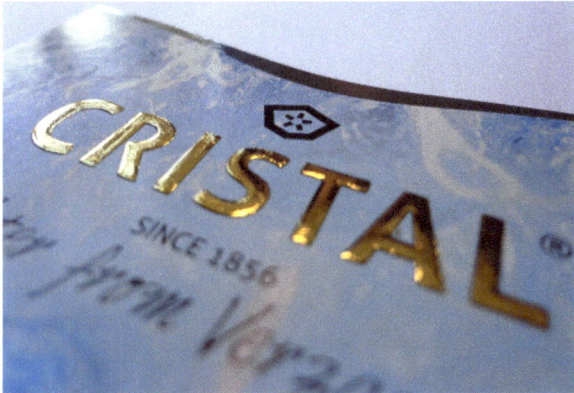

**Figure 6.8 -** Relief effect using metallic doming

## APPLICATION PROCESS

During metallic doming a thermo-reactive glue is applied using screen printing, then dried with UV light. The applied print image is then laminated with a thermo-active metallic foil. A soft, heated laminating cylinder presses the foil onto the glue. The foil wraps around the glue and, together with the glue, is heated by the laminating cylinder. This laminating process reactivates the glue and activates the foil.

This thermal activation means the foil sticks to the glue. After this, the printed image cools to reveal a metallic relief effect – this is the metallic doming effect (see Figure 6.9).

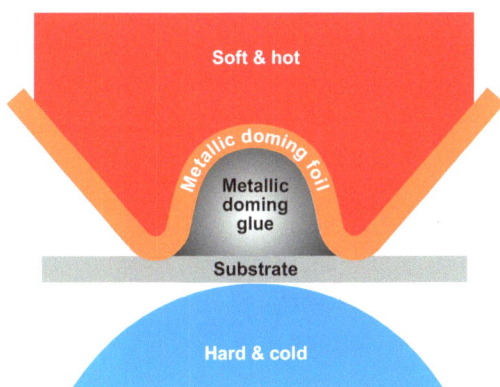

**Figure 6.9 -** The metallic doming process. Source: Gallus

## OVER-PRINTING METALIZED SUBSTRATES

Over-printing a translucent wash or varnish onto a metalized substrate is a good, low cost way to give a colored reflective effect similar to foiling or ink metalization.

The areas left unprinted on the label allow background metalized substrate to show through. Using a tinted translucent wash the metallic color can be easily adjusted. For example a yellow wash over a silver metalized substrate will give the appearance of gold or a translucent blue over silver will give a metallic blue (see Figure 6.10).

**Figure 6.10 -** Example of overprinting on metalized substrates

## OVERPRINTING FOILS

Overprinting ink onto hot and cold foil is also a trend worthy of note. In the past, foiling tended to be the last process, and on web presses the foil unit would have been at the end of the press. Now it is not uncommon for hot foil or cold foil units to be in the middle of the press in order to allow ink overprinting.

This allows nice effects to be achieved, but can pose a problem for the ink manufacturers and foil suppliers due to the poor ink receptivity of many grades of printed foil.

An interesting process from Color-Logic enables printers to achieve a 'hot foil look' by simply printing white and conventional CMYK inks over silver or holographic silver foils (Figure 6.11).

Using what is known as a Process Metallic Color System more than 250 metallic hues and special effects can be created. This technique eliminates the time-consuming and expensive hot stamping process, as well as the need to hold different foil colors in stock.

The Color-Logic system is compatible with offset, inkjet, flexography, digital presses, screen printing, and gravure processes.

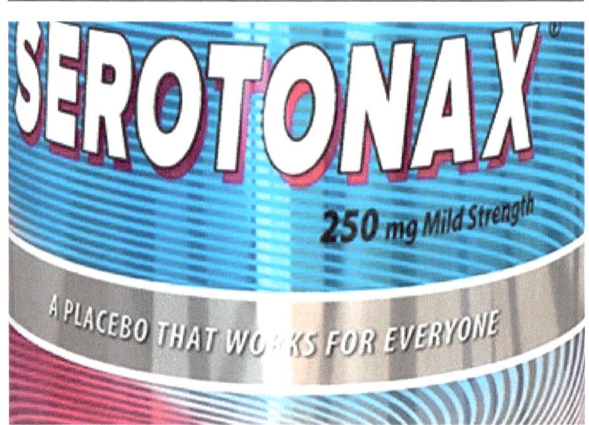

**Figure 6.11 -** Example of Color-Logic printed label

# Chapter 7

# Intaglio printing

Intaglio printing (which has become known as gravure printing) is a specialist print finish which is predominantly used for security applications.

Intaglio printing is the opposite of relief printing in which the image to be printed is etched or engraved below the surface of a print cylinder or plate. The printing process is carried out under pressure. The ink is consequently drawn up from the recesses onto the substrate and dries leaving a slightly raised image, which can be detected by feel.

## BRIEF HISTORY

Intaglio printing originated with the goldsmith engravers in about 1446. The images were hand engraved onto copper, gold and silver and the recessed image was filled with a black ink or enamel known as Niello and then pressed onto paper. These early prints were used by the goldsmiths to display the range of engravings available to the customer. The goldsmiths not only engraved their products, but also developed an etching method using nitric acid. In about 1640 a German engraver called Von Seigen employed a new method of intaglio printing called mezzotint, which was used to reproduce paintings in black and white and also in color. Engravers mastered the art of varying the depth of the engraving or etching which allowed differing shades of color to be achieved.

The intaglio process was further developed in the early seventeenth century when it became known as gravure printing. The process began to use metal plates which carried the etched image, which was then printed onto the substrate.

The invention of photography led to the method of transferring a photo image onto a carbon tissue coated in a light-sensitive gelatin. This process allowed the etching of an image onto a steel or copper covered rotary cylinder and heralded the beginning of the modern rotogravure process.

## THE PROCESS

The process of intaglio engraving and printing is relatively simple. The image to be reproduced is engraved onto a metal surface usually of copper, brass, zinc or steel. The image is hand engraved or etched and the lines/cuts hold the ink which is to be transferred onto the substrate.

The hand engraving of an intaglio image requires considerable craft skills. The engraver uses an engraving tool called a 'burin' which is a steel square ended tool which is diagonally shaped and sharpened at one end. This is used to cut a series of lines which can vary in both width and depth of the cut.

The deeper the line cut, the 'more' ink the line will hold and the shallower the line cut, the less ink the cut will hold. This effect produces the varying tonal values (see Figure 7.1).

## ETCHING

The other method of intaglio imaging is by the etching of the image onto the plate. The plate is covered in a film of acid-resisting wax called the 'ground'. The engraver uses an etching needle, to finely

Depressions are cut into a printing plate. The plate shown here is not to scale: the grooves can be fractions of a mm wide.

The plate is covered in ink

The ink is wiped off the surface of the plate, but remains in the grooves

Paper is placed on the plate and compressed, such as by a heavy roller

The paper is removed, and the ink has been transferred from the plate to the paper

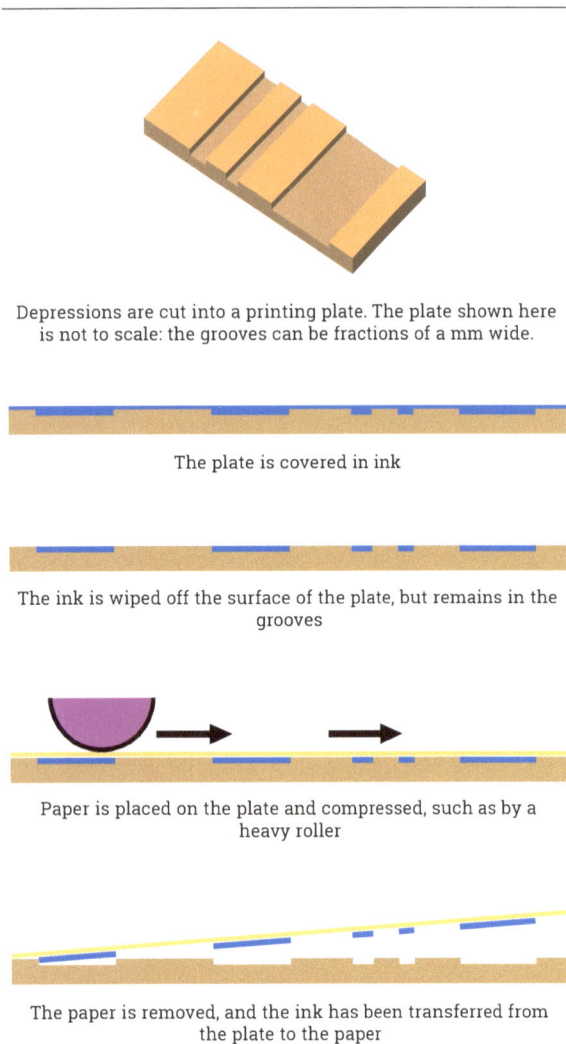

**Figure 7.1 -** The intaglio process explained

draw/engrave the image into the ground which in turn exposes the plate underneath.

The plate is then placed into an acid bath allowing the acid to etch into the surface of the plate where the ground has been removed thereby creating the recessed intaglio image. This is called the 'biting' process.

This etching process can be repeated by identifying the areas of the image that may require a deeper line depth. This involves applying a lacquer/varnish to these areas and re-etching with acid. This allows the engraver to vary the depth of the image to create the different tones of the image.

When the etched image has reached the required depth the plate is removed from the acid bath and the ground is wiped from the plate surface, which is now ready for printing.

## MARKETS

The use of the intaglio process in modern print relates very much to security printing, particularly in the printing of paper and synthetic substrates for currency, passports, banknotes and self adhesive postage stamps.

The processes used for the production of these complex products, often involves multi-process combination presses, equipped with lithography, flexography, screen and gravure/intaglio printing and a variety of finishing processes.

# Chapter 8

# Lamination

Film lamination is the process of applying a clear film to a printed web using a dry or wet adhesive to secure the film to the surface of the printed substrate.

Clear film is used as a means of enhancing the printed image and providing surface protection to a label during transportation and handling. It also provides a degree of moisture resistance, particularly for 'under the sink' and personal care products (Figure 8.1).

The lamination film can vary in thickness and clarity depending on the type of film used. This can be acetate, polyester or polypropylene based, all of which are supplied in reel or sheet format with either a matt or gloss finish.

Laminate film thickness is measured in microns. For instance 150 micron will be thicker than film at 75 microns.

Polypropylene or polyester laminates are suitable for application onto a printed web using an adhesive, combined with heat and pressure. The film requires careful handling and good control of the laminate tension during application is most important, ensuring that the laminate is applied smoothly and crease free onto the printed web.

Laminates are also supplied with a finish suitable for overprinting or as a non-printable surface. With products such as multi-layer labels, coupon and booklet labels a film laminate may be used to enhance, protect, hinge, and provide a reseal strip for the construction.

There are several types of lamination used in the packaging industry typically applied via a lamination unit (see Fig 8.2).

**Solventless lamination:** The adhesives used are

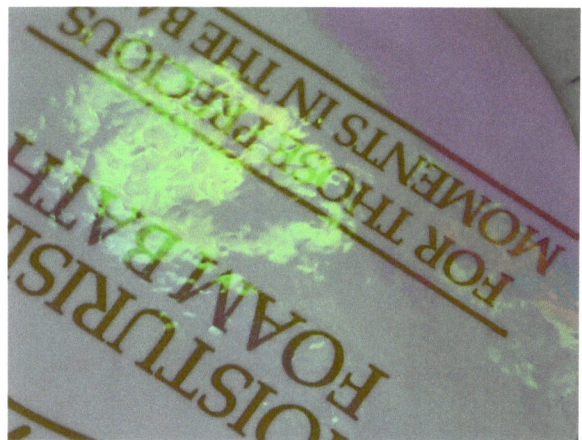

**Figure 8.1 -** Example of a film laminated label

solventless, which dry by chemical reaction and therefore do not requiring a drying system. This method is used widely in flexible packaging.

**Wax lamination:** The adhesive is a wax or hot melt which is applied in a liquid state to one of the substrates prior to the substrates being brought together. This method is used widely in the food packaging industry

**Dry lamination:** Is the process whereby an adhesive is applied to the laminate film, which remains slightly tacky before it is applied to the printed web. This process is used widely for self-adhesive label lamination (see Figure 8.3).

**Wet lamination:** With wet lamination the adhesive is in a liquid state when the laminate and base substrates are brought together. This process is used in both the flexible packaging and self-adhesive labeling industry.

## LAMINATING METHODS

In this Chapter the focus is on the two types of lamination most widely used in the self-adhesive label industry i.e. dry lamination and wet lamination.

### Dry Lamination

Dry lamination (described above) tends to be used for short run applications. The type of lamination film used for dry lamination is heavier and thicker than the film used for wet lamination and because it is pre-coated it is more expensive.

A health and safety issue associated with dry lamination is the amount of noise generated as the 'tacky' adhesive separates when unwinding from the reel.

Within the self-adhesive industry the adhesives used for dry lamination are generally water-based acrylics. Although some solvent-based adhesive is still used it is reducing due to the environmental issues created.

One of the interesting developments in laminate adhesives is the use of 100% solids adhesive. These are reactive, cured adhesives and their use eliminates the need for drying systems and the energy costs are significantly reduced.

With dry lamination there are two method of bonding the laminate film to the substrate being laminated.

## BONDING METHODS

Hot lamination uses a heated nip roller to activate a pre coated 'heat seal' adhesive on the laminate film. This process bonds the laminate film to the printed substrate.

In some instances the adhesive can be applied on the press as a hot melt application.

To apply the hot melt adhesive it is necessary to heat the hot melt to the correct temperature and then apply the liquid adhesive through a die coater unit. The two substrates then enter the nip roller assembly

**Figure 8.2 -** Typical lamination unit

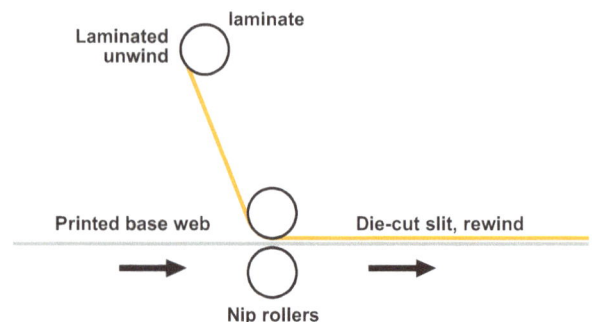

**Figure 8.3 -** Dry Lamination - Shows the layout of a dry lamination unit with the reel of pre-coated laminate positioned above the base web which is to be over-laminated

to apply the necessary pressure. The pressure of the nip rollers causes the adhesive to flow evenly, creating a consistent bonding between the two substrates.

Cold lamination uses a pressure sensitive adhesive which is pre-coated to the laminating film and remains in a 'tacky' state. The nip rollers are not heated, but do apply pressure to the laminating film and the substrate, bonding the laminate film to the printed base substrate.

This method of lamination avoids damaging the printed image.

## Wet lamination

With wet lamination a liquid adhesive is applied to the film laminate prior to the merging of the base web and the over-laminate.

In the process of wet lamination the adhesive is applied to either the reverse side of the film laminate or alternatively to the face of the printed substrate to be laminated. The liquid adhesive is applied by a roller coating system, usually the flexographic process. The two substrates, one of which has been coated with the adhesive, are then nipped to form the bond (see Fig 8.4)

## Types of wet adhesives

The types of adhesive used for wet lamination applications in the self-adhesive label manufacturing are;

- Water-based adhesives
- Solvent-based adhesives
- UV-based adhesives

The adhesive can be air dried using a heated drying unit to remove any solvent or moisture. If UV adhesives are being used, the web is passed through a UV drying system.

Water-based adhesives are considered better for bonding because of the good initial grab and fast drying characteristics, particularly if being applied to porous substrates.

Acrylic-based adhesives are a low cost liquid adhesive which provide excellent bonding properties and are suitable for UV curing. These UV cured laminating adhesives are solventless, low viscosity liquid adhesives, which can be applied through normal printing methods and then UV cured.

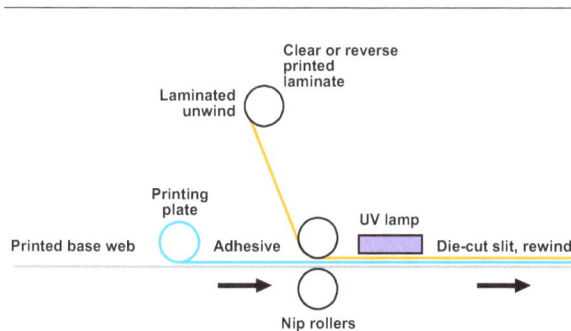

**Figure 8.4 -** Wet Lamination Process

# Chapter 9

# Thermography

Thermography is a heat activated print finish that in some forms can offer interesting opportunities to add a decorative quality to labels, packaging and other paper-based products.

Thermographic printing is a more cost effective and practical alternative to the die-stamping and intaglio printing processes. The difference between these processes is that the die stamped and intaglio process raises the surface of the 'substrate' (paper), whilst the thermographic process lifts the 'printed' image to create a raised 'embossed' effect on the image area.

There are two types of thermographic printing;
- The simplest method uses a substrate which has been coated with a heat sensitive material that changes color exposed to heat. This type of thermography when called direct thermal printing and was originally used in fax machines and shop receipt printers.
- The second thermographic printing method (which has more decorative qualities) works by conventionally printing the image and then coating the wet ink with a fine thermographic powder.

The printed and powdered image is exposed to a heat source and the ink and the powder fuse together which allows the powder to flow, forming a coating over the ink which is then cooled, allowing the coating to set and maintain the raised image. (see Figure 9.1)

Thermographic imaging is not recommended for heavy solids or fine text.

## THE PROCESS

The equipment used to produce the thermographic result consists of three units linked with a conveyor

**Figure 9.1 -** Thermography - close up of printed result

system (see Figure 9.2).

The first section is the printing operation which prints the image requiring thermographic processing. The press is usually a conventional litho, letterpress or screen printing press and the image is printed with slow-drying inks that do not contain drying agents and therefore remain open/wet.

The next stage of the process is the application of the thermosetting powder.

This unit applies the powder to the entire sheet and the powder sticks to the wet ink. The powder is delivered via a hopper which can be adjusted to alter the amount of powder being delivered to the sheet or web. The amount is governed by the size of the

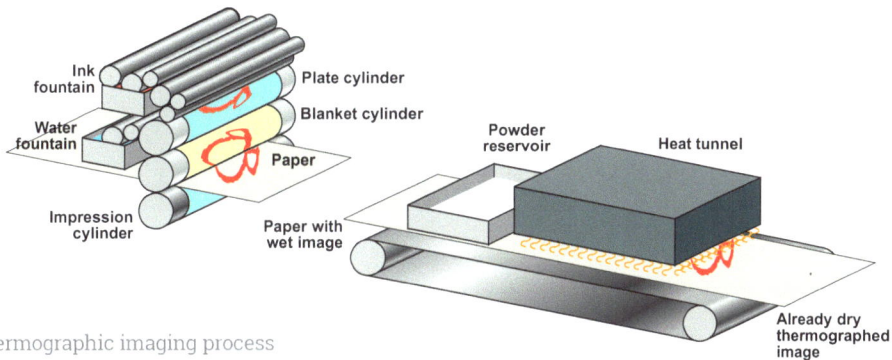

**Figure 9.2 -** Thermographic imaging process

printed image. For a small printed image less powder is required and more powder for a larger printed image. The excess powder is then removed using a low volume vacuuming system to extract all the powder from the non-imaged areas but leave the powder on the imaged areas.

The vacuum strength can be adjusted to ensure that the powder is not removed from the image area whilst the non-image area is powder free.

The thermosetting powder is manufactured from polymer resins. The most common polymers used are polyester and polyurethane. The process of manufacturing the powders involves mixing polymer granules with hardener and pigments. The mixture is heated in an extruder and then rolled flat, cooled and broken into fine chips which are roller milled into a fine powder.

The third part of the process is the application of heat to the sheet/web. The conveyor belt transfers the substrate into the heating unit where it is exposed to high temperatures for a period of approximately 3 seconds. The heat is conducted through the substrate, raising the temperature of the powder and the powder melts and fuses with the ink. It is important that the temperature is correctly set and maintained. The substrate then leaves the heater unit and is cooled, which solidifies the ink/powder creating the raised image.

## THERMOGRAPHIC SEQUENCE
- Image printed (ink remains wet and tacky)
- Thermographic powder applied to inked image
- Image exposed to heat (ink and powder fuse together)
- Image cooled leaving raised image with sharp edges

Some important points for achieving good thermographic results are identified below;
- Excessive heat will create irregular edges in the image, bubbling and pin holing
- Early melting of the powder during the heating process will cause the liquid powder to run over the edge of the image/type and distort the result
- The substrate cannot be exposed to a sudden intense heat as this induces considerable curl. A slower build-up of heat does not create as much substrate curl. A typical example would be to keep all the heater elements at a low temperature, bringing the powder up to the melt temperature much more slowly instead of a rapid high temperature exposure.

## ADVANTAGES AND DISADVANTAGES OF THE THERMOGRAPHIC PROCESS
### Advantages
- Low cost method of creating a relief image
- No dies or tooling required

### Disadvantages
- Slow process
- Exposing substrate to high temperatures

# Chapter 10

# Special effects inks, varnishes and coatings

There is a wide variety of specialty inks, varnishes and other special-effects technologies that can be used to add both decorative value and functionality to the surface of a label.

Special effects can add a new dimension to packaging, helping brands stand out from the crowd on the retail shelf. Special effect inks can also be interactive and thereby engage consumers on a different level, bringing unique functions as well as visual appeal into pack designs.

It should be noted that there are a wide range of inks and varnishes available that are able to perform dedicated brand protection tasks, adding layers of security to a product, protecting them against counterfeiting, theft or tampering. These specific security applications will be the subject of another book.

The focus of this Chapter will therefore be on inks and varnishes that have a decorative appeal, as well as perhaps a functional one.

## MATT AND GLOSS VARNISHES

Perhaps one of the simplest decorative effects that can be achieved is by printing gloss or matt varnishes onto the surface of a label

A varnish is a clear ink containing no coloring pigments. When printed or coated over the top of a printed or unprinted substrate in-line, the varnish gives a protective matt or gloss finish that enhances appearance and increases durability.

Varnishes are suitable for water-based, solvent or UV applications.

There are a number of varnish types available;
- Matt

- Gloss
- Semi-matt

Where the entire surface of the label is covered with a varnish, this is known as a 'flood coat'.

## SPOT VARNISH

Spot varnishes are typically printed in register to the label design. A clear matt or gloss varnish can be applied to specific areas of a printed label to improve the visual appeal of the label surface and in some cases provide surface protection, or improve slip or grip. The contrast between a matt and gloss varnish printed over a label surface offer interesting light refractive effects that can provide visual appeal (see Figure 10.1).

## LACQUERS

A lacquer is a clear resin/solvent coating, which may be glossy or matt, applied to a sheet or web of printed labels, to provide protection or resistance against scuffing, rubbing, chemicals, moisture, etc.

Carried out as an off-press operation, lacquering is performed on a machine equipped with a roller coater, after the inks have been dried. Since

**Figure 10.1 -** Patterned varnish effects – A gloss spot varnish is printed onto a matt varnish

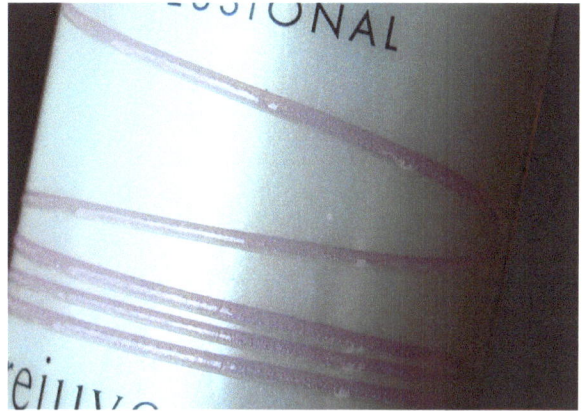

**Figure 10.2 -** Rotary screen raised image and tactile effects

lacquering is a separate operation, using expensive materials, it significantly increases the unit cost of the labels.

High performance UV lacquers are available in both matt and gloss.

It should be noted that glue-applied cut single labels, when they are printed using the gravure process, tend to use water-based or solvent-based lacquers. This process can however increase the tendency for the label to curl, which can result in problems during application to the container.

## TACTILE VARNISHES AND RAISED IMAGES - APPEALING TO A WIDER RANGE OF SENSES

Inks and varnishes can also be used to enhance the visual and sensory appeal of the label or pack.

Textures are becoming an important aspect of packaging design. New materials and processes offer designers opportunities to add a tactile quality to their packaging thereby appealing to a consumer's sense of touch.

Particularly popular are textured or raised surface effects. These are achieved by using special tactile inks that produce a relief effect similar to embossing (Figure 10.2).

The most popular way of achieving a texture finish is by using a special tactile ink delivered using the screen printing process. UV rotary screen is particularly effective as it can deliver up to 300 micron deposits in one pass (Figure 10.3).

**Figure 10.3 -** UV rotary screen printing

## BRAILLE

Textured or raised image varnishes can be used for printing Braille. Raised image tactile warning labels can be printed on a label surface to alert blind or partially sighted people to the potential dangers of the contents of drugs, medicines, hazardous chemicals, aerosols, etc. and distinguish them from harmless products.

Raised tactile labels for the blind have also been developed and used for wines and spirits (providing information on alcoholic strength), groceries and other consumer products.

Some countries have also introduced legislation (e.g. BS EN ISO 11683: 1997) that states that all labels sold to the public which are labeled 'very toxic', 'corrosive', 'toxic, 'harmful' or 'extremely flammable' must by law have a tactile danger warning label included, to alert the blind or partially sighted that they are handling a dangerous product.

These warning labels consist of a raised equilateral warning triangle or three raised dots forming the points of an equilateral triangle. The labels may be either incorporated into an existing label design or applied as a separate label (See Figure 10.5).

**Figure 10.5 -** Tactile Warning Triangle on labels with hazardous contents – used for the blind or visually impaired

For mass production of tactile warning labels on consumer products the label industry has developed a number of solutions, notably rotary screen printing units incorporated into a roll-label press can print deposits of up to 300 microns in thickness to provide the required tactile effects. Some converters are using a tactile varnish to print dots of precise dimensions (125 microns high and less than a millimeter in diameter).

## OTHER VARNISH EFFECTS

There are other varnishes which are being used that can add a tactile quality to the label surface, thereby appealing to the consumer's sense of touch.

**'Soft touch'** label varnishes (and materials) for example, are able to convey a warm velvety feel that can compliment a brand in certain markets. Soft touch varnishes are of low viscosity and therefore well suited for use on flexographic printing and coating units

**Grip varnishes** have a 'rubbery' feel and actually increase their grip potential when wet and have therefore been used in the beauty care sector on products such as shower gels.

## SCENTED PRINT - SCRATCH AND SNIFF

Appealing to the consumer's sense of smell, scented print or 'scratch and sniff' applications (as they are often referred to), add a new dimension to product promotion. In the toiletries sector for example, scented inks allow the user to pre-sample perfumes prior to purchase by scratching or rubbing a printed panel on the label to release the fragrance.

The preservation of fragrances on printed matter uses a technique called micro-encapsulation.

Fragrance capsules are mixed with an overprint varnish or ink and applied with an additional print unit on the press, so offering an economical solution for printing scents in combination with high quality printing.

Being encapsulated, the scent will not evaporate until the capsules are rubbed.

## PEARLESCENT DISPERSIONS

Inks or varnishes using pearlescent dispersions, based on the natural mineral mica, can create a unique lustrous effect on the surface of a label.

Mica particles or flakes are typically coated with a thin layer of metal oxide, such as titanium dioxide and/or iron oxide, to achieve a metallic sheen. The interplay of transparency, refraction and multiple reflections produces a spectrum of effects and colors - silver-white, gold and metal lustre (see Figure 10.6).

Pearlescent pigments produce their lustre through light being reflected, refracted and scattered when hitting the multiple surfaces of the pigment.

Due to particle size of the pigments, pearlescents are better suited to printing through a screen or flexographic unit. Pigments can be used alone or in combination with other colors.

**Figure 10.6 -** Pearlescent inks create a metallic lustre

## GLITTER INKS
Glitter Inks are made of multi-color metallic flakes suspended in a clear varnish or ink which can be can be applied to all or just part of a label design.

Glitter inks are available in a wide range of colors.

## IRIDESCENT INKS
Iridescent inks change color when viewed from different angles. The ink is colored with newly developed liquid-crystal polymers instead of conventional dies and pigments and is a mixture of two different liquid crystal components that can be cured with UV light.

The inks must be printed on top of another color

for the effect to be obtained; the best results being achieved when printing on a black background. Maximum effect is obtained when a heavy deposit of ink is used.

Applications are in the packaging of cosmetics, beverages and pharmaceuticals.

## THERMOCHROMIC AND PHOTOCHROMIC INKS
For information on thermochromic temperature change inks and photochromic light reactive inks. See Chapter 11, Smart, Active and Intelligent Labeling.

## FLUORESCENT INKS
The printed effect of using fluorescent inks is to add brightness and luminescence to the image printed on the label (See Figure 10.7).

The pigments in fluorescent inks absorb ultraviolet energy invisible to the human eye and then transmit them back as longer waves in the visible spectrum.

Fluorescent inks can be printed in either visible or invisible forms, which change color when subjected to shortwave light or long-wave light.

Historically these inks have been used for fraud protection, but are now used for promotional and other applications. The availability of pigments (short wave) for fraud protection is strictly controlled to preserve the authenticity of such inks. For some

**Figure 10.7 -** High impact graphics using fluorescent inks

applications Cyan, Magenta, and Yellow process inks can be replaced with their fluorescent equivalents to create high impact graphics.

Originally limited to screen printing, the pigments now have sufficient strength to permit colors to be printed in one impression by lithography, flexo and gravure.

## INVISIBLE FLUORESCENT INKS

These inks are invisible under normal lighting conditions, but glow in a variety of colors when exposed to UV light (Figure 10.8).

**Figure 10.8** - Invisible fluorescent inks – invisible message revealed under UV-A black light

## PHOSPHORESCENT INKS

Commonly known as 'glow in the dark' inks, phosphorescence occurs when a compound, after excitation ceases, continues to glow for a finite amount of time. This period of continued light emission or 'afterglow' will range from a few milliseconds to some hours.

This phenomenon can be used to characterize phosphors, or as a trigger in a machine readable system. They can also be used for promotional or decorative purposes. Items printed with photo-luminescent pigments have to be first charged in ambient light to allow them to then glow in the dark. The ink is transparent and can be printed by the

screen or flexographic process. Phosphorescent Inks will emit a green or in some cases a pale blue glow.

## OPTICALLY VARIABLE INKS

Optically variable inks (OVI) are inks which appear to change color when viewed from different angles. These inks go from a color to clear depending on the angle of the light.

Sometimes known as the 'flip-flop' type, OVI inks contain flakes of special film and are most effectively applied using a very heavy coating weight which can only be achieved using intaglio or screen printing.

There is a range of gravure and flexo printed optically variable inks which give reasonable effects.

OVI inks are expensive and are extremely difficult to replicate and are therefore used as an anti-reproduction device, for visual verification and protection against counterfeiting.

## SCRATCH OFF/RUB AND REVEAL INKS

Scratch-off inks are often used in 'scratch and reveal' type promotions or competitions where a hidden message or symbol is revealed (Figure 10.9)

An opaque coating or ink is applied to a specific area on printed matter to hide the details underneath. The hidden details are revealed when the coating/ink area is removed by scratching or rubbing with a fingernail or coin. Scratch-off inks are available in UV

**Figure 10.9** - Scratch Off Inks

and water-based formats and are typically printed over a release barrier which helps the removal of the opaque scratch-off layer.

In security applications, related ink systems are often referred to as 'Rub and Reveal'.

Patented ink security systems such as that on offer from Nocopi are impossible to replicate. Any paper-based document, tag, ticket or receipt can be secured using this versatile ink, which can be printed using the flexo process.

With 'Rub & Reveal' ink there is no need for special equipment to authenticate a document. After the ink is rubbed or scratched it uniquely changes colors thereby offering both physical and chemical verification.

## COIN REACTIVE INKS

With this type of reactive ink a coin is used to reveal a hidden image.

The pigment in the ink reacts with the metal of the coin to produce a grey color when scratched. A typical use would be for promotions where an invisible message is revealed when the consumer rubs an indicated panel with the edge of a coin (Figure 10.10).

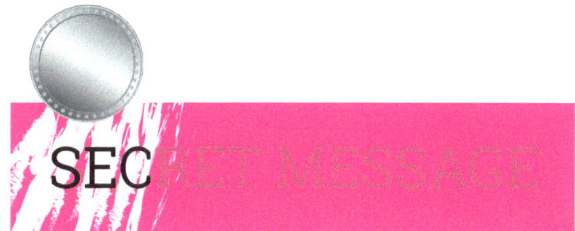

**Figure 10.10 -** Coin reactive inks

## SUMMARY

Find below a summary of the recommended processes for printing speciality inks. (Figure 10.11)

| PRODUCT | UV SCREEN | UV FLEXO | UV OFFSET | UV LETTERPRESS | WB FLEXO |
|---|---|---|---|---|---|
| Thermocromatic ink | ● ● ● | ● ● | ● | ● | ● ● |
| UV-Fluorescent ink | ● ● ● | ● ● ● | ● ● ● | ● ● ● | ● ● ● |
| IR-Fluorescent ink | ● ● ● | ● ● ● | ● ● ● | ● ● ● | - |
| Photochromic ink | ● ● ● | ● ● ● | - | - | ● ● ● |
| Phosphorescent ink | ● ● ● | ● ● | - | - | ● |
| OVS ink | ● ● ● | ● ● | - | - | ● ● |
| Coin ink | - | ● ● | ● ● | ● ● | ● ● ● |
| Scratch-off ink | - | ● | - | - | ● ● ● |

**Figure 10.10 -** Summary of the recommended processes for printing speciality inks. *Source: Flint Inks*
••• highly recommended  •• recommended  • available but not recommended  - not available

# Chapter 11

## Laser die-cutting

New developments in laser die-cutting are enabling end-users to add enhanced decorative value to their labels and packaging.

The ability of laser die-cutting to produce more intricate and complex cut-outs and to score and etch the substrate surface of the label or pack, will be explored in this Chapter.

---

Laser die-cutting will not eliminate conventional die-cutting, but complements it, and has a key role to play in the future of short-run, on-demand digital label converting.

It should be noted that the principles of conventional die-cutting are the focus of a dedicated Label Academy book on the subject.

### LASER DIE-CUTTING TECHNOLOGY

Unlike conventional die-cutting, laser die-cutting does not require a physical die. It can be described as the cutting out of a label or pack through the use of a computer controlled laser. The laser cutter uses programs developed from the step-and-repeat function of label origination to guide the cutting head around each label profile.

Laser cutters can be used off-line with re-registering of webs, or linked inline to digital or conventional label presses.

### EVOLUTION

The early progress of laser die-cutting in the label sector can be described as difficult. One of the early issues was the slow cutting speeds, a problem compounded on more complex shapes.

Another early issue involved the laser's inability to cut quickly around acute angles. In order to make acute turns the beam had to slow, allowing the heat of the beam to both penetrate the label facestock and more significantly the release liner. Burn through of the release liner has the potential to cause difficulties during the label application process.

Today's lasers are subject to greater control and can avoid the issue of liner burn-through by reducing the electrical power at the angle.

More specifically, developments in laser beam technology allow the beam to be emitted as a series of pulses with a focal point of several thousandths of an inch, and a cutting action generated from a series of overlapping dots. When the beam reaches a sharp corner, it emits fewer pulses at that point. The laser cuts through the face material, either paper or filmic and stops at the adhesive layer.

The digital nature of the laser cutter means that the die shape is derived from the pre-press digital design files, which means that there is virtually no limit to the range or frequency of shapes that can be cut.

### WHAT CAN BE ACHIEVED WITH LASER DIE-CUTTING?
#### • Complex profiles
Intricate and complex cut-outs not possible with conventional die-cutting are now possible using laser die-cutting (Fig 11.1). There are generally no size limitations. However more complex shapes will lower the web cutting speeds.

Lasers can cut most types of substrates, but it is not possible to cut PVC due to the toxic gases given off during the cutting process, or aluminum foil because its wavelength is too close to that of a $CO_2$ laser.

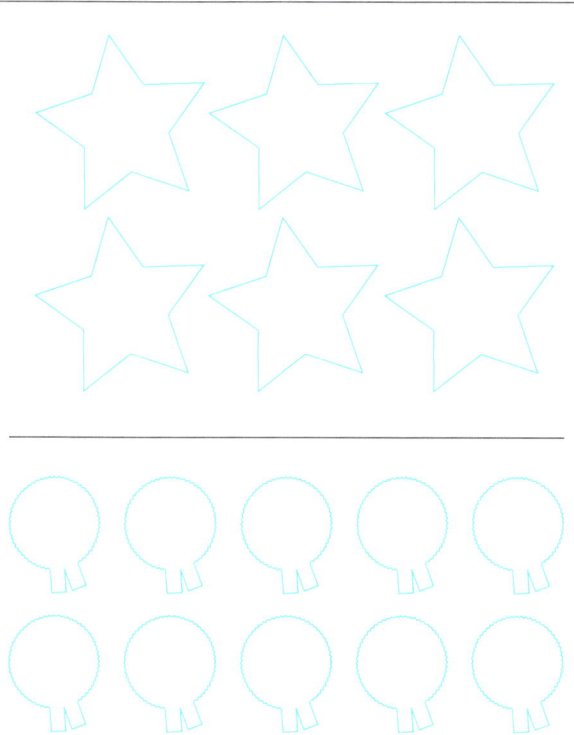

**Figure 11.1 -** Complex label profiles that are difficult to achieve using conventional die-cutting can be produced using laser die-cutting

- **Cut Out Areas**

Laser die-cutting offers opportunities to create intricate cut-out areas within the label or packaging design. Examples of laser die-cut labels and packaging can be viewed in Figures 11.2 a, b, c and d.

- **Personalization and brand protection**

Laser finishing can be used for surface etching for personalization and for coding. Laser systems can be used to etch OCR fonts, one and two-dimensional barcodes, serial/sequential numbers and codes.

The type of information that can be laser-etched is limitless and all in one pass with the cutting operation.

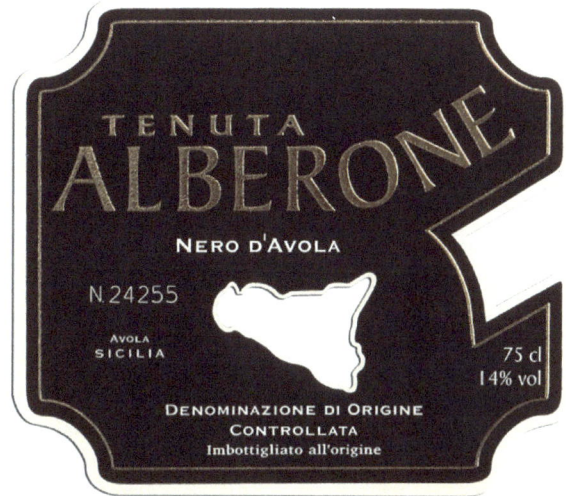

**Figs 11.2 a, b, c and d -** Complex cutout areas within the packaging design can be achieved with laser die-cutting.

- **Perforations**

Laser perforation of flexible packaging can create easy tear features on pouches and sachets.

- **Scoring**

Lasers can be used for scoring of folding cartons in-line or off-line with a digital press (Fig 11.3).

## KEY BENEFITS OF LASER DIE-CUTTING

Some of the key benefits of laser die-cutting are summarized below:

- More complex and intricate shapes and added value features are possible
- No dies are required, flat, rotary or flexible
- The die shape is generated from a digital file which means that there is virtually no limit to the range or frequency of shapes to be cut
- As no tooling is required turnaround is in minutes rather than days
- Represents a viable option on jobs where a large number of die and size changes are required
- Multiple depth cutting possibilities including kiss-cutting, thru-cutting and perforation in one pass
- Savings in set-up waste can be as high as 60 percent
- Offers significant reduction in labor costs

**Figure 11.3** - Laser scoring of cartons

# Chapter 12

# Smart, active and intelligent labeling

The focus of this book has been primarily on processes and techniques that are able to enhance the visual appeal of the label. New technologies are emerging however, that are extending the role of the label beyond that of the purely aesthetic and into one that is much more functional.

Interest in materials that react to environmental conditions such as light, heat, gases, pH and moisture is gathering pace. Some of the 'smart' and 'active' labels (SALs) have already found their way to market and many will emerge in the years to come adding a new dimension to packaging. There are

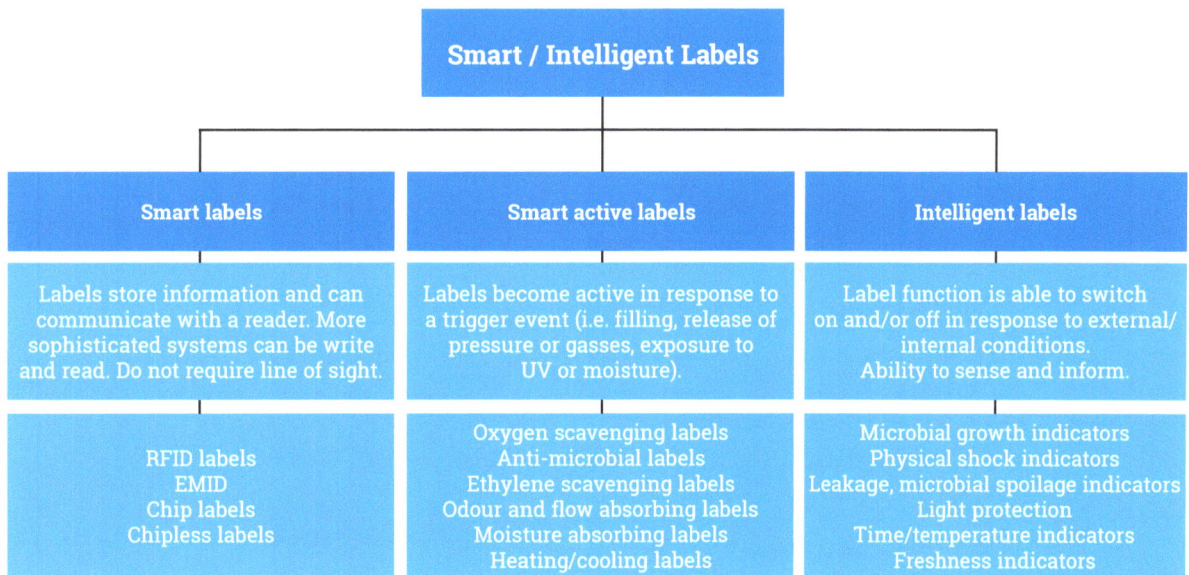

| Smart / Intelligent Labels | | |
|---|---|---|
| **Smart labels** | **Smart active labels** | **Intelligent labels** |
| Labels store information and can communicate with a reader. More sophisticated systems can be write and read. Do not require line of sight. | Labels become active in response to a trigger event (i.e. filling, release of pressure or gasses, exposure to UV or moisture). | Label function is able to switch on and/or off in response to external/internal conditions. Ability to sense and inform. |
| RFID labels<br>EMID<br>Chip labels<br>Chipless labels | Oxygen scavenging labels<br>Anti-microbial labels<br>Ethylene scavenging labels<br>Odour and flow absorbing labels<br>Moisture absorbing labels<br>Heating/cooling labels | Microbial growth indicators<br>Physical shock indicators<br>Leakage, microbial spoilage indicators<br>Light protection<br>Time/temperature indicators<br>Freshness indicators |

**Figure 12.1 -** The role and function of smart, smart active and intelligent labels. Source: *Encyclopedia of Labels and Label Technology*

many ways of creating SAL labels using smart inks, smart coatings, smart materials and by integrating components into the label structure.

Typically SALs are designed to sense the environment, it and to convey information to the user. The definitions between the terms tend to be somewhat blurred, but a common theme is the ability of these technologies to add a functional dimension to the label, so that it no longer remains a simple conveyer of brand identity. The ability to actively adapt to external influences means that the label can play a practical role in extending the shelf-life of a product, in promoting it, tracking it or protecting it from counterfeiters.

## MARKET DRIVERS

The growing interest in SAL packaging concepts in western markets is being encouraged by a number of drivers:

- Rising sales of convenience, fresh-food products that are placing increasing demands on the packaging industry for formats which act to preserve taste, appearance and nutritional qualities
- The demand for more product information by the consumer
- The need to improve product life tracking, communication and control within the packaging supply chain - particularly to indicate where product or packaging abuse has taken place
- Increasing pressure to reduce costs, and the desire to improve brand image and appeal

These demands, combined with recent advances in material science and biotechnology, are fuelling new applications in both food and non-food areas alike.

This Chapter aims to provide a brief overview of some of these emerging technologies.

## SMART COATINGS

The use of high-performance coatings can enable surfaces to be more responsive to their environment and add functionality. Smart coatings frequently rely on a visible color response to indicate a temperature change or respond to other stimuli, such as a pH change, oxidation, corrosion or electric current.

Applications for smart coatings include tamper evidence, time/temperature indicators, anti-counterfeiting solutions, anti-microbial properties, barrier performance and freshness indicators.

## 'ACTIVE' DEVELOPMENTS

Active packaging is designed to change the condition of the contents in order to extend shelf life or improve product safety.

The main focus of active packaging tends to be on maintaining or extending product shelf-life. The most commonly used 'active' technologies are desiccants, moisture absorbers and oxygen scavengers.

Whilst active packaging features tend to relate to the incorporation of additives into packaging film or within packaging containers, there is evidence that the label is being used for specific applications.

- ### Freshness Indicators

There are currently a limited number of freshness indicators on the market, where for instance an indicator label reacts to volatile amines from fish with a visual color change.

- ### Gas Indicators

Gas indicators attached inside the package can provide information on the integrity or correct gas concentration within the packaging headspace. For many perishable products exclusion of oxygen improves the stability of the product. A typical visual indicator can be formed as a printed layer or laminated in a polymer film and involves a distinct color change.

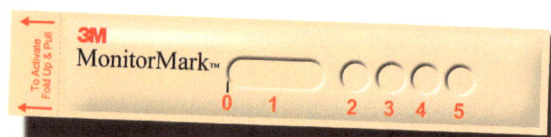

**Figure 12.3** - Typical time temperature tag.

• **Time Temperature Indicator (TTI)**

The time temperature indicator (TTI) is one area where the label is well established as a carrier (see Figure 12.3).

If perishable foods are stored above suggested storage temperatures, rapid microbial growth takes place and the product is spoiled. Typically attached as labels onto the pack surface, TTI's integrate the time temperature history of the packaging throughout the entire distribution chain and provide information about product quality.

TTI's play an important role in the healthcare sector. Maintaining the cold chain is fundamental to all vaccination programs, because vaccines and biological products are very sensitive to heat. A vaccine, for example, must remain active from its place of manufacture right through to the most remote areas where it is administered. If the cold chain is broken, the vaccine will lose its protective capacity.

## SMART INK TECHNOLOGIES

Development in smart ink technologies is extending the role of labels into new areas. Smart inks can be applied using most of the main printing processes. A summary of some of the main types are as follows;

• **Photochromic inks**

Photochromic inks exhibit spectacular color changes when they are illuminated with either ultra-violet light or natural daylight (Figure 12.4).

The inks are colorless in their natural state and when exposed they produce their visible color. These inks are reversible; reverting to their colorless state when the light source is removed. They are often used for product verification or in promotional applications.

• **Thermochromic inks**

Thermochromics are inks that change color with changes in temperature. They are available in two main types; reversible and irreversible.

Reversible thermochromic inks are generally used to show when the temperature of a pack or label has reached an optimum level. For example, when the temperature of a beverage (beer or wine) is at the appropriate level and ready to serve, the thermochromic ink label will change from white to blue. When the product warms up the message will disappear (Figures 12.5 and 12.6).

**Figure 12.5 -** Thermochromic inks that appear at pre-determined temperature thresholds can indicate that a product has reached an ideal serving temperature or play an important role in security labeling

**Figure 12.6 -** Thermochromic labels - on both these label designs the mountains turn blue when the beer is cool.

Irreversible thermochromic inks are used as time/temperature indicators in food and other packaging to provide a record of whether the products have been stored at the correct temperature, as well as in anti-counterfeiting or anti-tampering devices (Figure 12.7). It should be noted that the EU has recently issued a new standard stating that all thermochromic inks should not have over a recommended level of formaldehyde.

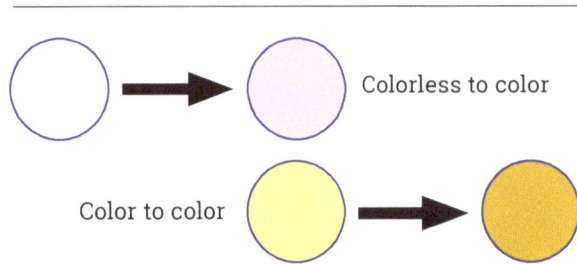

**Figure 12.7 -** Irreversible thermochromic ink changes

• **Heat and Reveal Inks**

Although thermochromic inks can be set at a variety of temperature thresholds, they are often set so that they can be activated at human body temperature, so that promotional messages or product verification can be achieved simply by touching the label or pack. In Figure 12.8 below black ink is printed over a 'hidden' image which is revealed when the ink is touched or rubbed.

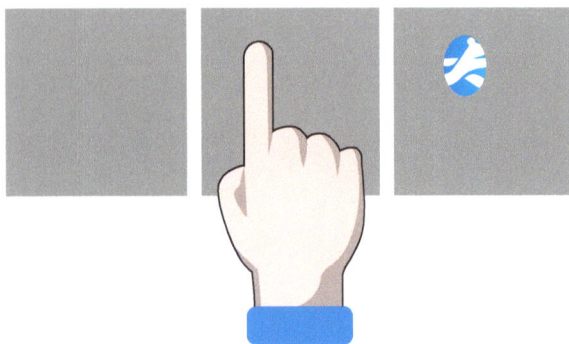

**Figure 12.8 -** Image revealed by normal body heat

## RADIO FREQUENCY LABELS

Labels that contain an RFID (Radio Frequency Identification) tag are often referred to as 'smart' because they can store information (such as a unique serial number) and communicate with a reader.

RFID labels are increasingly being used in a wide range of applications to replace product barcodes.

## WHAT IS RFID?

Radio Frequency Identification (RFID) is a term used for any device that can be sensed at a distance by radio frequencies. Not to be confused with EAS (Electronic Article Surveillance) tags found on many high value items such as CD's and DIY products, radio frequency identification permits electronic detection of identification numbers and other data held on the tag.

An RFID tag contains a microchip attached to an antenna that is able to pick up signals from and send signals to a reader. The tag contains a unique serial number and may also contain other information, for example, a customer's account number. Tag readers use the message they receive from the tag in accordance with their programer's instructions (Figure 12.9).

Generally, these readers must be no more than 2 meters (6 feet) away from the tags to be read. Unlike bar codes, RFID tags do not require line-of-sight reading, so a pallet load of RFID tagged goods can be read in a second or so.

Rather than a generic code, which relates to a specific product line – as with bar codes – each RFID tag has its own specific code. This offers retailers the opportunity of tracing and recalling certain batches of goods very quickly in the case of safety or quality concerns.

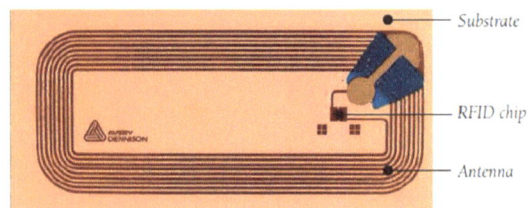

*The individual elements that make up an RFID inlay (13.56MHz). Diagram courtesy of Avery Dennison.*

**Figure 12.9 -** Typical structure of an RFID tag

## WHERE WILL IT BE USED?

It is clear that supply chain management will be the

principal future use for RFID with applications in inventory management, warehouse management, theft, out of stocks issues and transport and logistics.

RFID applications will be invaluable in improving inventory insight. Balancing supply, demand, and customer service requires real-time information and operational control that will be facilitated by this technology.

Items, racks, and shelves can be labeled to automate picking and put away with forklift-mounted and wearable printers. With RFID labeling routes can be identified and pallets, bins and shipping containers managed without human intervention.

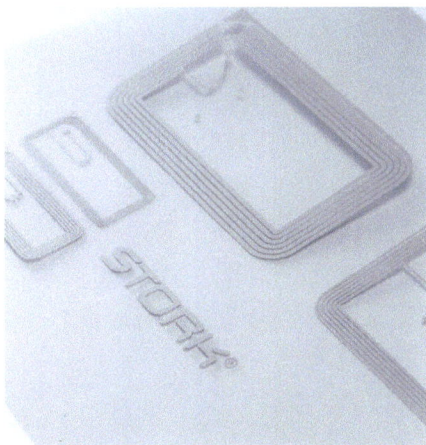

**Figure 12.10 -** SPGPrints collaborates with major suppliers of conductive inks to make RFID antennae

## PRINTED ANTENNAE

Antennae printed onto label structures using a conductive ink are increasingly being used (see Figure 12.10).

RFID antennae can be printed using silver conductive ink and there is work progressing on carbon graphite formulations.

In other instances smart inks contain radio frequency micro-taggants, which are responsive to electro-magnetic energy, enabling them to be detected when subject to an energy source. Information contained within the printed ink signature

on the label can be read remotely when a low powered energy source is aimed at the printed area.

Developments in low cost RFID are split into tags that have microchips and those that don't.

The more versatile chip tags comprise an antenna and a silicon chip mounted onto a substrate, but in the future it may be necessary to look towards 'chipless' RFID to reach the low unit prices required.

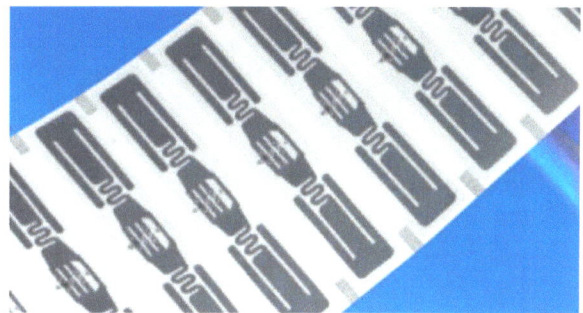

**Figure 12.11 -** A typical RFID inlay

**Figure 12.12 -** A typical RFID inlay

## INLAYS

Another route to integrating RFID into labels and packaging is to insert the complete device into the label structure itself in the form of an inlay (Figures 12.11 and 12.12).

RFID labels are typically produced by incorporating a chip and antenna on a thin polyester film into a roll label construction (Figure 12.13). The construction is then over-laminated with a print receptive face stock, printed, die-cut and applied as a standard self-adhesive label (Figure 12.14).

With the unit price per tag dropping and significant global players investing in RFID technology it is clear that low cost RFID is here to stay.

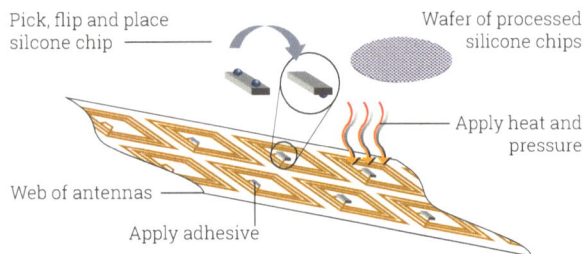

**Figure 12.13 -** Construction of a typical RFID roll label

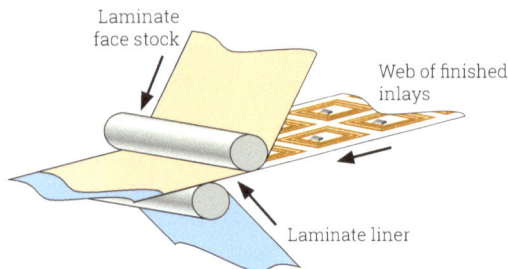

**Figure 12.14 -** Converting a web of finished inlays into a label laminate structure

## PRINTED ELECTRONICS

As we have seen with developments in printed RFID antennae, printed electronics are becoming an important facet of functional printing, creating a host of opportunities to add intelligent functions to a label.

Printed electronics involves printing conductive and dielectric inks on a flexible material to make an electrically active circuit or component.

When printing flexible electronics each printing process has pros and cons.

In the *Labels & Labeling* Yearbook 2015 Cal Poly's Dr Malcolm Keif outlines the pro and cons of each printing process in the production of a printed electronic circuit.

Screen printing is particularly good for laying down relatively thick ink films, which usually equates to better conductivity. However, screen printing is not ideal for fine features such as thin lines.

Flexo and gravure do better at reproducing small features, but have other challenges, including relatively thin ink films and inconsistent surface smoothness.

Inkjet printing is great for depositing fairly fine features and can make variable patterns, but it is quite slow compared to other processes.

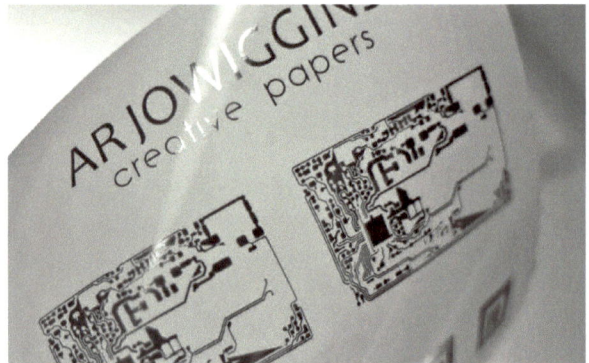

**Figure 12.15 -** Powercoat from Arjo Wiggins allows paper substrates to be used for printed electronic applications

Dr Malcolm Keif also details a list of printed electronics devices that can be reliably printed today.

- Conductive traces (think flexible wiring)
- Electrochromic displays
- Electroluminescent displays
- Batteries
- Organic solar cells (OPVs)
- Printed heating elements
- Antennae for RFID, NFC or any inductive coupling (such as charging a wireless toothbrush)
- Transparent conductive films
- Printed resistors, capacitors, mechanical switches, and other simple devices

Developments in this arena will extend the role of

the label from a static conveyor of brand values and information, into new realms where it can perform more complex functional tasks.

Paper substrates can be used for a variety of printed electronics applications (see Figure 12.15).

## OPPORTUNITIES FOR CONSUMER INTERACTION WITH PACKAGING

Coding printed on labels and packaging can offer a gateway to increased consumer interaction.

**Figure 12.16 -** Typical QR Code.

## QR CODES

Print can provide opportunities to build relationships with customers via QR codes printed on labels or packaging. Further information can be found in the *Label Academy Codes and Coding Technology book.*

QR (Quick Response) codes offer an instant way to get more information on a product by scanning the QR code with a smartphone and taking the user instantly to a web address (Figure 12.16).

An on-pack back label can only offer a limited amount of space for information but when a QR code and smart phone combine, access to a wealth of web based information can be unleashed. For example recipes, nutritional information or even wine tasting notes, and links to other related products, can give the consumer the benefit of making a more informed purchase.

Digital printing of codes can be done quickly and at a much lower cost than conventional processes.

## AUGMENTED REALITY

The introduction of new technologies such as the Smartphone and the trend of linking point of sale to online marketing resources is favoring more responsive and flexible print technologies.

Augmented Reality (AR) is an exciting new mechanism that is now being used to add value to packaging.

AR enables digital images/content triggered by a visual image to be seamlessly inserted into the real world through a smart phone, webcam or tablet. Brand marketers are bringing a spin to their packaging with a hidden level of graphics that can only be viewed through a phone's camera (Fig 12.17).

**Figure 12.17 -** Augmented Reality.

The technology can be added to any picture (even one that has already been printed) and it can also be programmed to allow the image to be scanned so it can be viewed afterwards, rather than having to hold the device over the image. This enables it to be used to demonstrate assembly instructions when building flat pack furniture, installing electronics or showing children how to build models.

The growth of smart phones and tablets will only increase the adoption of AR and there is currently talk of standardization in the industry to introduce a symbol that displays the presence of AR on packaging.

Current examples of AR packaging include drinks bottles that launch videos through an augmented AR smartphone app. and biscuit packs that use a QR code that directs the user to a video, photo or personalized message.

With little or no origination costs involved, digital print is a great way of reproducing the codes that link packaging at point of sale with web delivered content.

# Chapter 13

## Label leaflets and multi-layer constructions

Labels are assuming new functional formats, as retailers and manufacturers respond to demands for more information to appear on the product.

Increasingly popular are leaflets, booklets and multi-layer labels used for on-pack promotional coupons and to cater for extended text requirements.

This Chapter will outline the range of both multi-layer and label-leaflet constructions on the market and their method of manufacture.

It should be noted that many of the constructions detailed in this Chapter are protected by patents and have been the subject of much litigation over the years.

### PROMOTIONAL ON-PACK COUPONING

A coupon label can be defined as a label or part of a label, which can be used to supply information, or have a redeemable value. Coupons may be pressure-sensitive, non pressure-sensitive, dry-peel or adhesive blocked.

### TWIN-WEB CONSTRUCTIONS

Twin-web constructions are comprised of two separate web materials – usually a self-adhesive substrate and a non self-adhesive bond substrate – that are joined together to create a single web that performs two distinct functions. The base layer could, for example, perform the role of a standard product label whilst the top ply could be peeled away as a promotional coupon or even a receipt. It is possible to print on all the plies (see Figures 13.1 and 13.2).

The types of promotion which have been communicated by this method can be summarized as follows:

- On-pack advertising
- Money off schemes
- Cross promotion
- Competitions
- Fragrance (scratch and sniff) promotions
- Special Offers
- Recipes
- Collectables

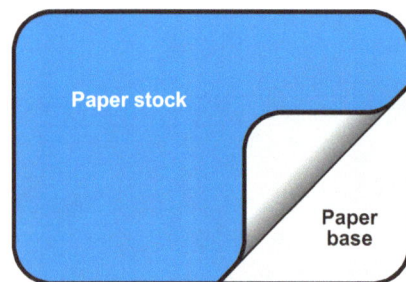

**Figure 13.1 -** Illustration of typical multi-layer coupon constructions

**Figure 13.2 -** Typical multi-layer booklet label

## COUPON MANUFACTURING

For self-adhesive coupon manufacturing the printing press requires a minimum of two web unwinds, each equipped with web guides. The self-adhesive web overlaps the bond substrate, so that registration is critically important. Good unwind tension control is required to ensure there are no 'wrinkled' glue joints.

The webs are printed front and back in the usual way and various combinations of adhesive and silicone are applied to the webs immediately before the lamination nip.

The choice of material can be critical, with both the bond substrate and pressure-sensitive face material being printable using the inks on the press, as well as by any subsequent variable print method chosen. The substrates should have good lay-flat, and sufficient strength to withstand die-cutting, matrix stripping, perforation, hole punching, fan folding, etc.

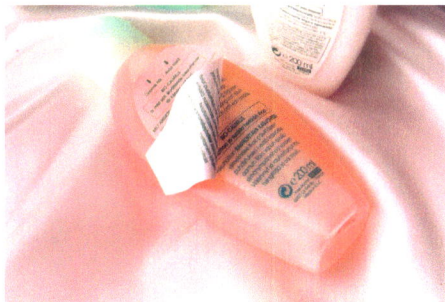

**Figure 13.3 -** Two ply peelable coupon

In use, the top layer of the construction should be easily removed by the consumer as a 'tack-free' coupon (Figure 13.3).

It is possible to add hinges and re-seal strips using hot melt adhesive to form simple re-sealable booklets. Also, by using a clear filmic base layer, peelable coupons can be created that leave no trace when applied to packaging (i.e. the packaging graphics show through the clear carrier).

Some coupons can be manufactured in-line on presses fitted with what is called a 'web-shift' mechanism. This allows the printed web to be slit lengthways down the web. The two narrower webs are then placed one above the other using the 'web-shift' mechanism to create the multi-layer structure.

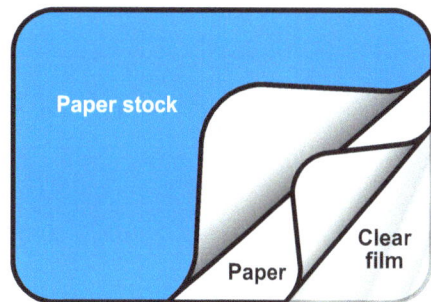

**Figure 13.4 -** Typical 3 ply coupon construction with 5 printable panels

## MULTI-WEB CONSTRUCTIONS

In some cases more than two webs can be brought together to enable multiple coupons to be produced (see Figure 13.4). Clearly more printing heads and unwinds/rewinds will be required to facilitate the manufacture of these type of constructions.

## SANDWICH CONSTRUCTIONS ('PIGGY-BACK LABELS')

'Sandwich' labels consist of a double-layer label material whereby the top layer of the label can be removed and used as a 'peel and apply' sticker.

Whilst normal self-adhesive label laminates are produced from two plies, the so-called sandwich (or 'piggy-back') construction has been developed, most commonly of three plies. They have two label layers and one liner. The top layer has an adhesive on the back whilst the second layer has a release coating on the face and an adhesive coating on the reverse (see Figure 13.5).

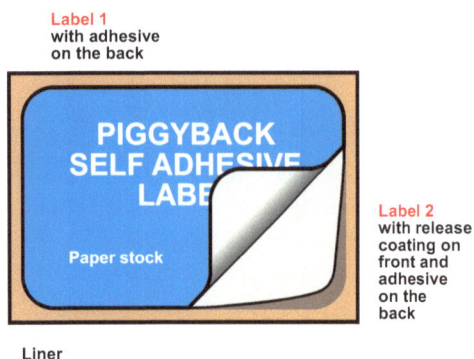

**Figure 13.5 -** Diagram of piggyback self-adhesive construction

**The construction is die-cut down to the third ply.**
When applied, the second ply, which can be self-contained carbonless paper, translucent glassine or clear film, has an adhesive coating which adheres the piggy-back label to the labeling surface, leaving the top ply as a peel-off label for subsequent use by the end user.

**BOOKLET LABELS**
Booklet labels combine a self-adhesive label and a pre-printed leaflet in a single construction. These extended text leaflets have been developed to provide a means of adding information to a pack.

There are now a wide range of booklet labels, leaflet labels, fold-out labels, fold-in labels, concertina labels and other extended text multi-page label formats (see Figure 13.6). They are widely used in the agrochemical, DIY, pharmaceutical, chemical and promotional labeling sectors for information, instruction and marketing applications.

The demand for label-leaflets and booklets is being driven by a number of factors;

- The rise in self-medication products
- Tighter legislation relating to packaging information
- Increasing demand for clearer labeling of products, particularly ingredients and safety information
- Packaging minimization – the desire to reduce loose leaflet-carton combinations and other bulky packaging

Typical uses for multiple page booklet labels include:
- Assembly and handling instructions
- Conditions of sale
- Cooking/processing/recipe details
- Copyright, patent and trade mark information
- Diagrams to aid understanding
- Directions for use (including dilution or mixing tables)
- Disclaimers
- Dosage instructions
- Environmental, Health and Safety statements
- Hints and tips
- Mixing instructions
- Multi-language information
- Statutory warnings
- Technical diagrams
- Warranty information

**PHARMACEUTICAL LEAFLETING**
Security is the essential product identity need of the pharmaceutical industry and the self-adhesive booklet label satisfies the need for all-important information to stay with the product.

Using a self-adhesive booklet-label ensures that;
- The correct information leaflet is issued with the product
- The content of the information leaflet is legible and correct
- The patient receives the information and is able to use the product as specified

The life and death standards, vital for both the health of the patient and the consequent commercial health of the industry, have powered many technical innovations in the labeling and leafleting businesses.

## MANUFACTURING EXTENDED TEXT BOOKLETS

Extended text and multi-page booklet/leaflet labels can be produced in a wide variety of formats and designs, to suit virtually any container type. They can be produced in any number of colors, with glossy or matt varnishing or transparent lamination to protect the information. Extended text booklets are often permanently fixed to the container or pack, but perforated or peel-off removable options are available. A re-seal capability can be specified if required to allow the booklet to be sealed after opening (see Figure 13.7).

There are three main methods of producing a booklet label

**Figure 13.6 -** Extended text leaflets solve the problem of overcrowded labels

### • In-line – one stage

With this method booklet labels are produced in-line on a narrow-web label press that is capable of accommodating a second web. The self-adhesive base web in printed as a normal label. At the same time the second non-adhesive web is printed on both sides (the web is turned over using a turner bar) before being folded using a plough folder. The folded web and base carrier web are brought together in register and secured in place, typically, with a filmic over-laminate. The booklet labels are then die-cut and wound into a reel ready for automatic application (see Figures 13.7 and 13.8).

A wide variety of permanent and removable pressure-sensitive adhesives are available.

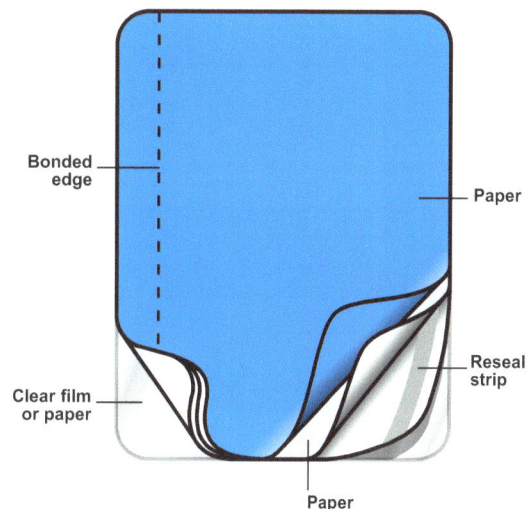

**Figure 13.7 -** Typical resealable booklet construction on a self-adhesive carrier (produced in-line)

### • In-line – two stage

Using a two stage manufacturing method the leaflet element of the booklet is printed and folded off-line.

The leaflets are then stacked into a dispenser (or 'on-serter') which is mounted onto a narrow web label press. The leaflets are then applied in register to a self-adhesive carrier web. To ensure consistent application the 'on-sert' unit is electronically linked to the press drive shaft.

The leaflet is held in place on the carrier web by static electricity or an adhesive before it is over-laminated and die-cut.

Other manufacturing variations use silicone or adhesive application in patches, stripes or patterns, with a wide variety of folding and finishing solutions.

The finished booklets are wound onto reels ready for automatic dispensing.

### • Off-line – multi-stage

With this manufacturing method concertina and multipage leaflets and booklets are pre-printed - usually by sheet-fed offset

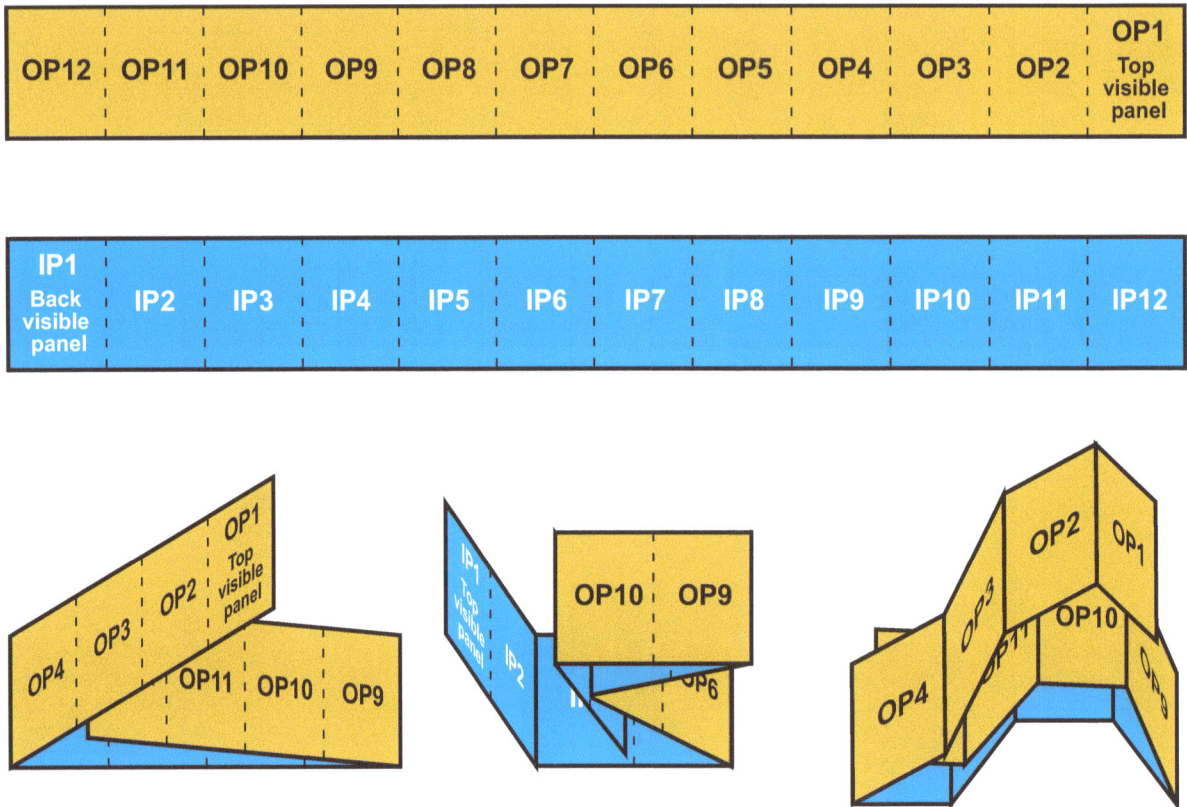

**Figure 13.8 -** Diagram of typical booklet label construction. The booklet is folded in-line before being glued or laminated to a self-adhesive base layer

The leaflets are then folded before being combined with a self-adhesive label on specialist equipment.

They are die-cut and supplied on reel for use with standard label application equipment.

The number of pages for booklet labels can vary, with the maximum usually 96 pages.

Apart from providing additional space for descriptive information, warning symbols and diagrams, multi-page solutions can incorporate bar codes, holograms and tamper-evident options, or can be overprinted with batch numbers, sequential numbers, date or use-by information, manufacturer codes or machine-readable data.

**REVERSE SIDE PRINTING (BACK PRINTING)**
Reverse printing involves printing on to the underside of a transparent film, with the printed image then read from the front side of the film. The printing plate(s) need to have the image areas reversed so that the image can be viewed correctly through the pack (Figure 13.9).

This is achieved by de-laminating a self-adhesive laminate on a suitable label press, printing onto the adhesive and then re-laminating the roll before further printing and die-cutting.

If this technique utilizes a clear label material, reverse printing offers a completely protected print image that is scratchproof, with the finish of the label

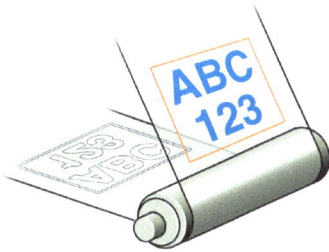

**Figure 13.9 -** Printing on the underside of a transparent film with the printed image then read from the front

**Figure 13.10 -** Reverse printing on filmic materials – the image reverse printed on a clear back label is viewed through the pack and clear contents

film providing a glossy or matt surface to match the container.

This effect is commonly used in the toiletries and cosmetics sector to create an innovative decorative effect where the pack graphics on the clear container are viewed through the product itself where the liquid is clear or transparent (Figure 13.10).

A typical reverse printed label involves the following steps;

- The image is printed in reverse on the underside of a transparent film
- The image is overprinted with an opaque masking layer (high opacity inks applied by rotary screen or UV flexo printing are often used)
- Additional graphics in the normal orientation are printed on top of the opaque mask layer.

This type of printing allows the reverse printed image to be visible through the pack whilst the label surface can be used as a standard back label. Printing on both sides of the label (face and adhesive) saves using another label.

# Index

www.ingramcontent.com/pod-product-compliance
Lightning Source LLC
Chambersburg PA
CBHW041723210326
41598CB00007B/754